Biologically-Inspired Systems

Volume 7

Series Editor
Prof. Dr. Stanislav N. Gorb, Christian Albrecht University of Kiel, Kiel, Germany

More information about this series at http://www.springer.com/series/8430

Jonas O. Wolff • Stanislav N. Gorb

Attachment Structures and Adhesive Secretions in Arachnids

Springer

Jonas O. Wolff
Functional Morphology and Biomechanics
Zoological Institute
Kiel University
Kiel, Germany

Department of Biological Sciences
Macquarie University
Sydney, Australia

Stanislav N. Gorb
Functional Morphology and Biomechanics
Zoological Institute
Kiel University
Kiel, Germany

Supplementary videos to this book can be accessed at http://www.springerimages.com/videos/978-3-319-45712-3

ISSN 2211-0593 ISSN 2211-0607 (electronic)
Biologically-Inspired Systems
ISBN 978-3-319-45712-3 ISBN 978-3-319-45713-0 (eBook)
DOI 10.1007/978-3-319-45713-0

Library of Congress Control Number: 2016957511

Printed on acid-free paper

This Springer imprint is published by Springer Nature
The registered company is Springer International Publishing AG
The registered company address is: Gewerbestrasse 11, 6330 Cham, Switzerland

Preface

In 1974, Nachtigall, one of the pioneers of biomimetics, compiled the first review on means of attachment in nature, aiming to be understandable for both biologists and engineers (*Biological Mechanisms of Attachment*). Sixteen years later, one of us followed up on this exciting approach and compiled the first comprehensive overview on attachment structures in arthropods, with a particular focus on insects (*Attachment Devices of Insect Cuticle*). Later, the books *Biological Adhesives* edited by Smith and Callow (2007) and *Biological Adhesive Systems* edited by von Byern and Grunwald (2010) provided an excellent overview on biological adhesive secretions.

With this book, we focus on the second largest class of animals, belonging to some of the most diverse and most abundant organisms on earth, but to date highly neglected in terms of bioadhesion: the arachnids, which include, among others, spiders, mites, harvestmen and scorpions. Arachnids differ in many regards from other arthropods, like insects and crustaceans, and spider silk is not the only example for the evolution of high-performance biological materials in the spotlight of biomimetics and bioengineering. Our book both reviews current literature on attachment devices and adhesive secretions in arachnids and presents novel, previously unpublished data, in order to provide a comprehensive (but by no means complete) overview on this subject.

Our studies would not have been possible without the continuous support by numerous peers and private persons, supplying material and/or photographs and bringing in their expertise and skills: Siegfried Huber (Mühlhofen); Michael Seiter (University of Natural Resources and Life Sciences, Vienna); Dr. Axel Schönhofer (Nackenheim); Esther Appel, Dr. Lars Heepe, Achim Oesert, Fabienne Frost, Jessica Nahrstedt, Ingo Grawe, Marina Wirth, Kirstin Dening, Theresa Gödel, André Karstedt, Chantal Wiegmann, Neele Ott and Nadine Sydow (University of Kiel); Dr. Peter Jäger and Julia Altmann (Senckenberg Research Institute Frankfurt); Prof. Gerd Alberti, Dr. Peter Michalik, Dr. Katrin Kunz, Melanie Witthuhn and Dr. Anja Klann (University of Greifswald); Prof. Jochen Martens (Johannes Gutenberg University

Mainz); Prof. Wolfgang Nentwig, Dr. Lucia Kuhn-Nentwig and Benjamin Eggs (University of Bern); Prof. Jutta Schneider, Tomma Dirks and Angelika Taebel-Hellwig (University of Hamburg); Rainer Foelix (Neue Kantonsschule Aarau); Martín Ramírez and Ivan Magalhães (Museo Argentino de Ciencias Naturales); Facundo Labarque and Darrell Ubick (California Academy of Sciences); Prof. Rosemary Gillespie (University of California, Berkeley); Dr. Christopher Taylor (Curtin University, Perth); Hay Wijnhoven (Nijmegen); Prof. Glauco Machado and Solimary García-Hernandez (University of São Paulo); Dr. Daria Bajerlein (Adam Mickiewicz University); Dr. Steffen Bayer (State Museum of Natural History, Karlsruhe); Pedro Coelho (CIBIO Research Center in Biodiversity and Genetic Resources); Dr. Suresh Benjamin (Zoological Research Museum Alexander Koenig, Bonn); Arno Grabolle (Weimar); Michael Schäfer (Berlin); Jürgen Guttenberger (Lauterhofen); Sylvia Voss (Borkwalde); Carsten Wieczorrek (Köln); Jörg Pageler (Oldenburg); Salvatore Canu (Usini); Tobias Bellmann (on behalf of his deceased father Dr. Heiko Bellmann); Dr. Walter Pfliegler; Arthur Anker; Markus Gottlieb; Tobias Töpfer; Lennart Bendixen; and Martin Lemke. We would like to use the opportunity to express our warm thanks to all of them.

We sincerely thank the German National Merit Foundation (Studienstiftung des Deutschen Volkes) for the PhD scholarship and Macquarie University for the research fellowship to J.O. Wolff, making this study possible. Additionally, this work has been supported by various DFG (German Science Foundation) and BMBF (Federal Ministry of Science and Education, Germany) grants to S.N. Gorb.

Special thanks to Springer Publishers, especially Zuzana Bernhart and Mariska van der Stigchel, for their professional guidance at different stages of the manuscript preparation.

Kiel, Germany

Jonas O. Wolff
Stanislav N. Gorb

Structure and function of biological systems as inspiration for technical developments

Throughout evolution, organisms have evolved an immense variety of materials, structures, and systems. This book series deals with topics related to structure-function relationships in diverse biological systems and shows how knowledge from biology can be used for technical developments (bio-inspiration, biomimetics).

Contents

Chapter 1
Overview

Abstract Attachment is one of the major types of interactions between an organism and its environment. There are numerous studies that deal with this phenomenon in lizards, frogs, insects, barnacles, mussels and echinoderms, but the second largest class of arthropods, the Arachnida, is highly neglected. This book surveys the attachment organs and structures, and adhesive secretions occurring in this class of animals and discusses the relationship between morphology and function, evolutionary trends, and biomimetic potential. In this chapter we give a brief introduction into the topic, describe the methodologies used, provide a unifying terminology of attachment organs in arachnids, and give an overview on the attachment principles and their distribution among arachnid orders.

1.1 Introduction

Attachment is a fundamental part of the interaction between an organism and its environment. In dependence of the phylogenetic origin and the ecological demands a huge variety of attachment organs and structures evolved. They are based on different fundamental physical principles like friction, dry and wet adhesion, cohesion, mechanical interlocking, suction and penetration (Nachtigall 1974; Gorb 2001; Smith and Callow 2007). Recent efforts in resolving the functional principles of attachment structures in insects and geckoes at the macroscopical, microscopical and molecular scale led to intense interdisciplinary research and, not at least, technical innovations, like the recently developed Gecko® Tape (Binder GmbH) (for review see Gorb 2011). However, the second largest class of arthropods, the Arachnida, has been highly neglected in this research. A comprehensive overview on attachment devices in arachnids has never been undertaken, and there are only few studies that deal with the mechanical function of some of these systems.

Beside the structure-based attachment devices, a large diversity of gluey secretions can be found among all kinds of organisms. Glues are based on various substances ranging from polysaccharides and lipids to polypeptides and proteins (Betz and Kölsch 2004; Smith and Callow 2007; von Byern and Grunwald 2010). Depending on their composition and biological function, the bonding to the substrate can be temporary or permanent. Functions of biological glues include attach-

J.O. Wolff, S.N. Gorb, *Attachment Structures and Adhesive Secretions in Arachnids*, Biologically-Inspired Systems 7,
DOI 10.1007/978-3-319-45713-0_1

ment to the substrate, prey capture, nest and cocoon construction, defense, camouflage, and reproduction. Studying the mechanical function of such secretions, their efficacy and limitations, may help to understand the biology of these animals and their ecological relationships with the environment. Furthermore, many of these natural glues exhibit a set of properties that surpasses that of artificial adhesives in several aspects. Often with small amounts they can create strong bonds even on repellent or wet surfaces, under water, or with living tissue. They are bio-degradable, use non-toxic solvents and might be bio-compatible (Smith and Callow 2007). Hence, they might have very innovative potential for the development of technical and medical adhesives. Furthermore, understanding the function and restrictions of biological adhesives may help to develop efficient counter-measures against parasites and pests, like ticks, varroa mites or spider mites.

This work aims to explore the structural and mechanical principles of cuticular attachment devices and secretion products in different orders of Arachnida LAMARCK 1815. It focuses on the morphology, mechanical properties and biological functions of these structures, sets them in relation to life history, ecology and phylogeny of the studied animals, and discusses their potential for future biomimetic approaches.

In the following, we give a short introduction to arachnids, describe the methodology used to obtain the original data presented in the book, provide a list of definitions for specific scientific terms, and present an overview on the attachment principles and their distribution among arachnids.

1.2 The Arachnid Orders

1.2.1 Spiders (Araneae)

Spiders are, besides some mite orders, the most diverse and abundant order of arachnids, well-known by everybody. To date they comprise about 45,800 species (World Spider Catalog 2016) in 114 families and occupy most terrestrial habitats on earth, including sub-arctic regions, deserts, shores and in a single case, the water spider *Argyroneta aquatica*, even permanently submerged in water. Spiders are well recognizable by a couple of unique characters: a constriction between cephalothorax (in spiders called prosoma) and abdomen (called opisthosoma), silk glands and spinnerets at the end of the opisthosoma, jackknife-like chelicerae with venom glands, and the modification of the pedipalps into a copulatory organ in males. The building of webs is commonly attributed to spiders, however, at least half of all species do not use silken traps to capture prey, but hunt it down themselves. The history of spider-like arachnids reaches back to the Devonian, since there are fossils of spinnerets from that period (Shear et al. 1989; Selden et al. 2008). True spiders are sub-divided into three sub-orders: the burrowing Mesothelae, regarded as living fossils showing a segmented opisthosoma, the Mygalomorphae, which are bird spiders and allies, and the Araneomorphae, the largest sub-order comprising all modern spiders, including orb weavers and hunting spiders.

1.2.2 *Scorpions (Scorpiones)*

Scorpions are well recognized by their crab-like chelae on the pedipalps and their tail (metasoma) equipped with a venom gland and a stinger. They are one of the larger arachnids, and widespread all over the world, with the exception of the cooler regions. Currently there are about 2220 known species of scorpions (Rein 2015). Most of them live at the ground, some of them are burrowers, other hide under stones. However, they may also occur in shrubs and on trees, where they rest under loose bark.

1.2.3 *Harvestmen (Opiliones)*

Harvestmen or daddy-longlegs are the third largest order of arachnids, with about 6620 species worldwide (Kury 2015). They are especially known from the temperate zone of the Northern hemisphere, where they may be very abundant and noticed for their very long and thin legs in comparison to the compact globular body. The legs are detached when gripped by a predator and often keep on moving after the separation from the body. The name 'harvestman' refers to this phenomenon, because the movement of the separated leg reminds of a sickle cutting wheat. However, there are many harvestmen that are short-legged, most of which live on the ground or in the litter layer. The highest diversity of harvestmen is found in the tropics. The tropical species are often armored with spines, and might be vividly colored, thus having a rather different appearance as the temperate zone harvestmen. In contrast to most other arachnids, which are strict predators, harvestmen frequently feed on dead invertebrates, fungi and plant material, in addition to captured prey.

1.2.4 *Whip-Spiders (Amblypygi)*

Whip-spiders and whip-scorpions (see below) are small orders of arachnids that are closely related to spiders. They are nocturnal hunters, distributed in the tropics and subtropics. Whip-spiders exhibit a flattened body with tilted (latigrade) legs, very long and thin feeler-like front legs and long raptorial pedipalps. There are about 190 species of whip-spiders living at damp places, preferably in caves, or in tropical rain forests at the ground and under the loose bark of old trees (Seiter et al. 2015). Whip-spiders can reach remarkable body sizes, but due to their cryptic lifestyle they are usually not encountered. However, some of the larger species have been frequently shown in movies, zoos, and introduced into pet trade in the recent years, due to their impressive habitus and behaviors.

1.2.5 Whip-Scorpions (Uropygi)

Whip-scorpions (super-order Uropygi) comprise two orders, the large epigaeic Thelyphonida (also called vinegaroons), and the small hypogaeic Schizomida, both of which are relatively similar in their morphology despite of their different size (Kästner 1941). The whip-scorpions share the characters of feeler-like front legs and strong raptorial pedipalps with the whip-spiders. Besides they exhibit an antenna-like appendage at the end of the abdomen, the flagellum, which is rather long in thelyphonids, but often short, and sometimes plate-like in schizomids. There are about 100 thelyphonid species and 220 schizomid species, all of which are distributed in the tropical and sub-tropical regions (with the exception of some establishments of schizomid populations in European greenhouses) (Harvey 2003). Vinegaroons might be quite abundant in some areas, coming out at night hunting on the ground. They spread acids when disturbed. Schizomids, though more diverse, are usually not noticed, because most of them live in the soil and are rather small (body length of less than 5 mm).

1.2.6 Camel Spiders (Solifugae)

Camel spiders are fast running ground hunters with a cylindrical body and strong scissor-like jaws. They can reach enormous body sizes (up to 15 cm). With few exceptions, they are nocturnal desert dwellers, and may withstand extreme heat and drought (Cloudsley-Thompson 1961). There are currently 1100 known species, distributed in Africa, America, Asia and Southern Europe (Bird et al. 2015).

1.2.7 Pseudoscorpions (Pseudoscorpiones)

Pseudoscorpions are small, mainly soil and litter dwelling arachnids, hunting for mites and other small soil arthropods. On the first view, pseudoscorpions look like miniature scorpions, with large pincer-like chelae on the pedipalps, however, they lack a tail and stinger. Nonetheless, many pseudoscorpions have a venom gland, which is located in the pedipalpal chela. Further, they may be able to spin silk from a spinneret on the chelicerae (Weygoldt 1969). Some pseudoscorpions are known from bee hives, where they represent an effective control against the bee parasite-*Varroa destructor* (Donovan and Paul 2005). Pseudoscorpions are distributed all over the world, with the exception of the arctic regions. Currently about 3440 species are known (Blick and Harvey 2011).

1.2.8 *Mites and Ticks (Acari)*

The Acari are the most diverse group of arachnids, both in terms of ecology and morphology. It is a presumably paraphyletic super-group comprised of two super-orders, the Parasitiformes (=Anactinotrichida), and the Acariformes (=Actinotrichida). Despite of the name both super-orders contain both free living and parasitic lineages. The Parasitiformes comprise the orders Opilioacarida (=Notostigmata), Holothyrida (=Tetrastigmata), Mesostigmata (=Gamasida) and Ixodida (=Metastigmata). The Acariformes comprise the Trombidiformes (=Actinedida) and Sarcoptiformes.

The Opilioacarida is a small order of ground-dwelling mites, with some basal character, most noteworthy the segmentation of the abdomen, which is lost in all other mites and ticks. Thus they resemble small harvestmen, leading to the name of the group. The Holothyrida are regarded as a relict of tick ancestors, which are not parasitic yet, but feed on the body fluids of dead arthropods. Ixodida are the well-known parasitic ticks feeding on the blood of vertebrates. The large order Mesostigmata comprise mainly predatory mites feeding on small arthropods. Some of them are efficient means of pest control in agriculture to defeat spider mites or aphids. Others are negatively recognized as pests, such as the bee parasite *Varroa destructor*.

The Trombidiformes are perhaps the most diverse mite order in terms of life-styles and body shapes, having conquered nearly all habitats from high mountains to the deep sea. They comprise predatory mites, as well as parasites and plant sap suckers, like gall mites (Eriophyidae) and spider mites (Tetranychidae), which are important pests. Well-known are the red velvet mites (Trombidiidae), which are often remarked during gardening, and with the African *Dinothrombium* species including the largest mites (up to 12 mm). Some lineages have exploited both fresh-water and salt water, with some species living in highly salty or hot springs. The order Sarcoptiformes is divided in two major groups, the free living Oribatida and the predominantly parasitic Astigmatina. The Oribatida are an important part of the soil fauna feeding on detritus breaking down the organic matter, but they may also be found on trees, where they feed on algae and lichens. Most of them are of a globular shape with a capsule-like armor. The Astigmatina are primarily tiny and unpigmented vertebrate parasites, such as feather mites (Psoroptidia) or scabies mites (*Sarcoptes scabiei*).

Due to the large diversity of the acarids and the estimated great number of un-described species the compilation in this book is far from being complete, but will summarize the significant diversity of attachment organs and sticky substances of this animal group.

1.2.9 Hooded Tickspiders (Ricinulei)

The Ricinulei are the smallest order of arachnids, with 76 known species (Fernández and Giribet 2015). They are restricted to the tropical regions of Africa and America, where they occur in the soil and in caves. The hooded tickspiders have a flat body, often a reddish color, and a unique hood-like hinged plate, called cucullus, which covers the mouth parts.

1.2.10 Micro-Whipscorpions (Palpigradi)

Palpigrades are the most cryptic order of arachnids, living deep in the soil or in cave systems. They lack any pigmentation and are tiny (<3 mm). Their general habitus resembles that of uropygids, with a long hairy flagellum. However, their pedipalps are not modified into raptorial legs, but unusually used for walking, whereas the first leg pair is elongated and used for sensing. Recently, it was observed that palpigrades feed on cyanobacteria, which are scraped off the substrate with the pincer-like chelicerae (Smrž et al. 2013). Currently there are 85 described species of palpigrades (Giribet et al. 2014).

1.3 Material and Methods

1.3.1 Literature Survey and Material Origin

Morphological and functional studies on attachment devices in arachnids are rare. To find hints on the distribution of such organs, structures or the use of adhesive secretions, zoological monographs and descriptive taxonomical literature was assessed (see citations in the text).

During visits of the arachnological collections of the Senckenberg Research Institute Frankfurt and the Zoological Museum at the University of Greifswald conserved material of different arachnid orders was studied by means of a dissecting microscope to find specific attachment organs.

On the basis of this preliminary survey the taxa of interest were selected and both living individuals and specimens conserved in ethanol were acquired for morphological and functional studies. The material for these studies was collected in the surroundings of Kiel, during a conference travel to Taiwan and a collection trip to the South-Western Alps. Living individuals were also obtained from breeding stocks from the Institute of Ecology and Evolution at the University of Bern, the Ethology Group at the University of Hamburg and private persons. Further specimens were obtained from private collections and the arachnological collections of the Senckenberg Research Institute Frankfurt and the Zoological Museum at the University of Greifswald.

We studied living individuals of spiders (Araneae: Mesothelae, Mygalomorphae and Araneomorphae), whip spiders (Amblypygi: Pulvillata and Apulvillata), whip scorpions (Thelyphonida and Schizomida), harvestmen (Opiliones: Laniatores, Eupnoi and Dyspnoi), scorpions (Scorpiones), pseudoscorpions (Pseudoscorpiones), ticks (Ixodida) and mites (Mesostigmata, Trombidiformes and Sarcoptiformes). Conserved material was obtained for all arachnid orders except Holothyrida and Palpigradi. See Appendix A2. for a list of species from which original data was obtained.

1.3.2 Light Microscopy

Light microscopical images were recorded with a multifocus stereo microscope (Leica M205 A, Leica Microsystems GmbH, Wetzlar, Germany) equipped with a camera (Leica DFC420), and an inverted microscope (AXIO Observer.A1, Carl Zeiss AG, Oberkochen, Germany) with a mounted B/W high speed video camera (Fastcam SA 1.1, Photron Inc., San Diego, CA, USA).

1.3.3 Scanning Electron Microscopy (SEM)

Standard SEM For Standard SEM material stored in 70 % ethanol was used. The specimens, which were whole animals or animal parts, were dehydrated in a series of increasing ethanol concentrations (80 %, 90 %, 100 % and 100 % on molecular sieve), and critical point dried. Dried samples were glued on stubs using a carbon-rich sticky tape and sputter coated with 10 nm Au-Pd. Specimens were studied with a Hitachi S 4800 scanning electron microscope (Hitachi Ltd., Tokio, Japan) at an acceleration voltage of 3.0 kV.

Cryo-SEM Living individuals (most of investigated mites, some pseudoscorpions, scorpion prenymphs and whip scorpion prenymphs) were attached to a sample holder using Tissue-Tek® compound and shock frozen in liquid nitrogen. The frozen samples were sputter-coated with 10 nm Au-Pd using the Gatan ALTO-2500 cryo system (Gatan Inc., Abingdon, UK) and viewed in the Hitachi S 4800 SEM with the stage cooled down to −120 °C. Some specimens were slightly touched with a frozen scalpel mounted in the prechamber in order to produce freeze fractures.

Histological Sections Living individuals were anaesthetized with carbon dioxide and legs were clipped off using micro scissors. Legs were immediately fixed with buffered 2.5 % glutaraldehyde and 1 % osmium tetroxide, dehydrated in a series of increasing ethanol concentrations and embedded in Epon. Epon blocks were trimmed with a Leica EM TRIM2, and sections of 1.0 and 1.5 μm were made using a Leica EM UC7 ultra-microtome with a Diatome MC2391 diamond knife (Diatome

AG, Biel, Switzerland) to about a half of the embedded specimen foot. The rest was cut from the block with a fine circular saw mounted onto a drilling machine and further processed like the sections. Some sections were placed on SuperFrost™ glass slides for light microscopical investigations. The others were placed on cover slides (previously cleaned with absolute ethanol and 70% acetone and coated with pioloform). Sections attached to the slides were bathed for 45 min in a concentrated KOH solution in two parts of methanol and 1 part of propylene oxide (Maxwell 1978). Similar procedure for the half cut block samples lasted for 1.5 h. Sections were air dried, whereas the block samples were put in absolute ethanol and critical point dried. Further treatment and microscopy were as described above (Standard SEM).

1.3.4 High Speed Videography (HSV)

Standard HSV For observations of pretarsal kinematics, living individuals running on horizontal and vertical glass slides were filmed. High speed videos were taken using a color high speed video camera (Fastcam 1024 PCI, Photron Inc., San Diego, CA, USA) mounted onto a stereo microscope, which could be positioned vertically, horizontally and upside-down. To obtain high frame rates, a bright light source was focused onto the foot and a white sheet of paper was held behind. Videos were taken with post-action trigger and rates of 250–3000 frames per second.

Reflection-Interference Contrast Microscopic HSV (RICM-HSV) For investigation of the actual contact area of adhesive foot pads, its change during locomotion and the deposition of fluid secretions, living individuals running on thin glass slides were studied by means of an inverted microscope (AXIO Observer.A1, Carl Zeiss AG, Oberkochen, Germany) with a mounted B/W high speed video camera (Fastcam SA 1.1, Photron Inc., San Diego, CA, USA) and operated in reflection interference contrast mode (coaxial light). Videos of detaching feet were taken at rates of 1000–8000 frames per second and a shutter time of 1/10,000 s, using post-action trigger.

1.3.5 Pull-off Force Measurements

Adhesive forces to smooth glass slides were measured with living individuals of whip spiders (*Sarax brachydactylus,Charinus cubensis*, and different stages of *Charon* cf. *grayi*), pseudoscorpions (*Neobisium* sp.), scorpion prenymphs (*Liocheles australasiae*) and spiders (*Zoropsis spinimana*, females). With the means of non-toxic two compound Polyvinylsiloxane (PVS) dental wax the animals were glued onto a thin wire mounted onto the cantilever of a micro force transducer (FORT-10 with 10 g force range, World Precision Instruments Inc., Sarasota, FL, USA). The force transducer could be moved vertically by a micromanipulator at a speed of

0.25–0.50 mm per second. The signal of the force transducer was amplified and processed by a Biopac MP-100 acquisition system (Biopac Systems Ltd, Goleta, CA, USA). Force curves were recorded using the AcqKnowledge 3.7.0 software (Biopac Systems Ltd). Each animal was tested at least at three different days with 10–25 measurements in total. Peak forces were divided by animal weight (measured with an AG 204 Delta Range scale, Mettler Toledo GmbH, Greifensee, Switzerland) to calculate the safety factor (sf), and by the sum of contact areas of all legs in contact to calculate the adhesive strength. In order to estimate the effective contact area, single feet were filmed with means of RICM-HSV (see above) during the test. The contact area at highest forces (usually before contact breakage) was used for further calculations. The contact area was measured as the proportion of the pad being in direct contact (indicated by dark contrast in the image), not the actual contact area given by microstructures and fluids. Otherwise the adhesive strength could hardly be compared between animals. Further, to get a better estimate of the effective contact area under natural load, the contact area of each foot in each individual, when hanging upside down on a glass slide, was measured. For this purpose, images were recorded in a stereo microscope (Leica M205 A, Leica Microsystems GmbH, Wetzlar, Germany) equipped with a camera (Leica DFC420) and coaxial illumination.

1.4 Terminology

In arachnids, no consistent terminology is present for their adhesive organs and pretarsal structures. Hence we first summarize the termini used in arachnological literature and attempt to uniform them following the definitions established in entomology by Holway (1935) and Dashman (1953). Etymology follows Maggenti et al. (2005).

***Ambulacrum* (also *ambulacral apparatus*)**. From Latin = ambulare = to walk. The ‘foot’ of ticks and mites (Acari). It usually refers to all structures distal of the tarsus. It is usually the last eudesmatic segment (=leg segment connected to muscles) with all its appendages, and hence may be homologous with the arthropod pretarsus. In some mites the basal pretarsus is subdivided and can be highly elongated and equipped with enclosing sclerites (*ambulacral stalk*). Distally, there is the section that makes the effective contact with the ground, often including a ventral arolium or empodium and up to three claws (*ambulacral claws*). In ectoparasitic mites, the claws may be reduced and the arolium greatly enlarged, forming a round disc (*ambulacral disc*).

Apotele From Greek *apo* = away from + *telos* = end. In mites the distal most appendages, including the tarsal claws, empodium and arolium. Some authors, in contrast, apply this term to the ambulacral stalk. It is presumably an appendage or part of the pretarsus. However, some authors call the part distally to the ambulacral stalk the pretarsus. Hence comparative approaches are necessary to homologize this structure. Dunlop (2002) applied this term more general to all chelicerates, naming the pretarsus with all its apical structures, although he was unsure about the homology. In these cases, the term *apotele* should be avoided in favour of the traditional term *pretarsus*.

***Arolium* (pl.: arolia)** From Greek *arole* = protection. Median un-paired lobe- or cushion-like attachment organs situated at the pretarsus are called *arolia*. They arise from the median pretarsal section and may contain sclerotized elements. They may occasionally be apically sub-divided.

Caruncle From Latin *caro* = flesh. Soft, bloated pretarsus in acariform mites, forming a 'fleshly pad' for substrate attachment. This term is synonymous to the term *arolium* and should be replaced accordingly.

Claw Tuft Hairy adhesive pads in the distal leg of spiders (Araneae). The structure is very similar to the *scopula* (see below), but the term is well-established to clearly distinguish between the leg pads and the foot pads.

Empodium From Greek *en* = in + *pous* = foot. In insects, the *empodium* is an outgrowth of the *unguitractor* (a sclerite between the claw retractor tendon and the claws). It can be claw-, bristle- or pad-like. We use this term for an unpaired sclerotized protuberance of the ventral pretarsus, usually in the shape of a little claw. In trombidiform mites it can also be a hairy adhesive foot pad in the shape of a flat lobe covered with spatulate microtrichia ('*tenent empodium*'), or it may be modified into an arolium, like in pseudoscorpions.

Onychium From Greek *onyx* = little claw, nail. In insects, it is occasionally used synonymous with the *arolium*. In spiders, it is occasionally used synonymous with the *empodium*; or to name a distal sub-segmentation of the tarsus, as in Scytodidae, Drymusidae, Oonopidae and Orsolobidae. Due to its very different usage, the term is confusing and imprecise and should be replaced by respective synonyms.

Pretarsus From Latin *prae* = before + Greek *tarsos* = foot. The pretarsus is the most distal leg segment which is controlled by muscles. In insects, there is only a claw depressor muscle, whose tendon inserts at a ventral sclerite, which bears the claws (the *unguitractor plate*). In arachnids, the (synapomorphic) pretarsus is reduced to a thick ring- or bow-like sclerite (*pretarsal sclerite*), controlled by a ventral *depressor (protractor)* and a dorsal *levator (retractor)* muscle (de Meijere 1901).

***Pulvillus* (pl.: pulvilli)** From Latin *pulvillus* = little cushion. In insects, the term traditionally describes paired lobes underneath the claws. They are often the outgrowths of the claw articulation membranes and not the median pretarsal membranes. This term is divergently applied on the median foot pads of ticks and amblypygids and should consequently be replaced by the term *arolium* (see above). In mites, this term is also occasionally used for large tenent setae underneath the claws (Vitzthum 1931), which is misleading.

Scopula From Latin *scopa* = broom (Homann 1957). This term is traditionally used for a dense brush of modified setae in arachnids. The term has been established for pads of spatulate (*tenent*) setae on the legs of Araneae and Opiliones. In spiders a distally separated section of the scopula underneath the claws is called *claw tuft* (see above).

Spatula From Latin *spatula* = spoon. A flattened and broadened apical tip of a *tenent seta* or of microtrichia on a tenent seta.

Tenent Plate *Tenent* from Latin *tenere* = to hold. The basal sclerite of a *claw tuft* (see above), also *claw tuft plate*. It is highly sclerotized and carries a dense array of sockets, in which the tenent setae are situated. The tenent plate is part of the tarsus.

Tenent Seta Latin *seta* = bristle. Setae are hair- or bristle-like cuticle protuberances originating from a single cell and usually joined to the underlying cuticle with an articulated socket, which originates from another cell (Beutel et al. 2013). Tenent setae, also called *adhesive setae*, *scopula setae* or *aliform setae*, serve attachment purposes. They are characterized by a spatulate tip or by the possession of spatulate microtrichia (see *spatula*).

***Unguis* (pl. *ungues*)** Latin = claw. The *pretarsal claw*, often erroneously called *tarsal claw*. The claw is a usually paired hook-like structure apically on the pretarsus, articulated by basal arthrodial membranes. The claws can be smooth or pectinate. One or both claws can be reduced. The median, often claw-like *empodium* is not an unguis, as it is an outgrowth of the pretarsal sclerite and not articulated with it. In some trombidiform mites, the claws can be equipped with spatulate microtrichia ('*tenent claws*').

1.5 Distribution of Attachment Principles Among Arachnids

Like insects, arachnids are terrestrial arthropods of often small size. Accordingly the demands of attachment and climbing through complex microhabitats are comparable. Many arachnids are predators of, primarily, other arthropods and hence must have means of grasping or trapping their prey. Others evolved a parasitic lifestyle, in particular mites and ticks (orders Mesostigmata, Ixodida, Trombidiformes, Sarcoptiformes), which means a high necessity to stay in place on the host. Also for reproduction purposes, such as indirect sperm transfer via a spermatophore or for egg clutch placement, attachment plays an important role in most arachnids. Hence, it is not a surprise that we find an enormous diversity of attachment devices in arachnids, with striking analogies, on one hand, and unique features, not found in any other group of animals, on the other hand.

The types of attachment devices can be categorized according to different functional principles and structural features (Fig. 1.1), which will be discussed in the subsequent chapters in detail. The found diversity comprises interlocking and clamping devices (claws, spines, hooks, pincers, expansion anchors and raptorial legs), structural adhesives (smooth and hairy adhesive pads), suckers, and hardening or viscid glues.

The distribution of functional principles among Arachnida is summarized in Table 1.1. Mechanical attachment is found in every arachnid order. The smooth adhesive pads, which are based on a softening of the cuticle occur in pseudoscorpi-

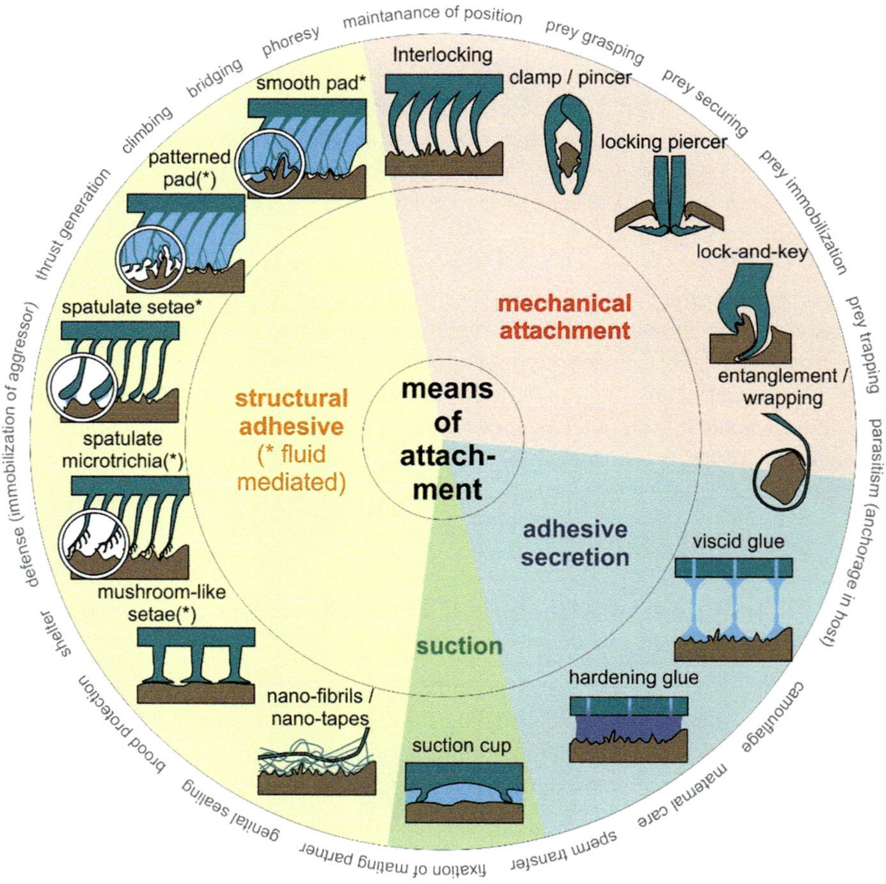

Fig. 1.1 Schematic overview on different mechanisms and biological functions of attachment occurring among arachnids

Table 1.1 Summary on attachment devices in arachnids

		Araneae	Amblypygi	Uropygi	Opiliones	Scorpiones	Pseudo-scorpiones	Solifugae	Acari	Ricinulei
Mechanical attachment	*Interlocking*	2 claws (+ empodium);	2 claws;	2 claws + empo-dium; spines; spermatophore head	1–2 claws (+ empodium); spines; setal sole; penal barbs	2 claws + empodium; spines	2 claws;	2 claws; spines	1–3 claws;	2 claws;
		Serrated bristles; frictional setae; spines;	Spines;				Spermato-phore head		Spines (rare);	Copulatory legs
		Some mating plugs	Spermato-phore head						Body microstructures of peristaltic Nematalycidae	
	Clamp/pincer	Chelicerae;	Chelicerae; raptorial pedipalps	Chelicerae; raptorial pedipalps	Chelicerae;	Chelicerae; pedipalp chelae	Chelicerae; pedipalp chelae	Chelicerae	Some chelicerae; clamping legs; raptorial legs; whole body	Chelicerae; pedipalp pincer; cucullus
		Whole-body capture basket; raptorial leg in *Trogloraptor*			(raptorial) pedipalps					
	Expansion anchor	Chelicerae	–	–	–	–	–	–	Tick hypostome; some chelicerae	–
	Lock-and-key	Male pedipalp + epigyne	–	–	?	–	–	–	Sucker + stick in some Acarida	?
	Entanglement/wrapping	Cribellate silk;	–	–	Prehensile tarsi	–	–	–	Silk	–
		Aciniform silk (bridging line and prey wrapping)								

(continued)

Table 1.1 (continued)

		Araneae	Amblypygi	Uropygi	Opiliones	Scorpiones	Pseudo-scorpiones	Solifugae	Acari	Ricinulei
Structural adhesive[a]	*Smooth pad*	Arolium in *Nopsides* spp.	Prenymphal arolium	Prenymphal arolium	Nymphal arolium	Prenymphal arolium	Arolium	Pedipalp organ; arolium	Arolium; anal pads of Heterozerconidae	–
	Patterned pad	–	Arolium	–	–	–	–	–	–	–
	Spatulate setae	–	–	–	Scopula	–	–	–	–	Scopula
	Spatulate microtrichia	Scopulae; Claw tufts	–	–	–	–	–	–	Tenent setae, empodium and/or claws	–
	Mushroom-like cuticular devices	–	–	–	–	–	–	–	'sucker'- setae of Acaroidea and Nematalycidae; some arolia	–
	Ribbon-like contact elements	Tape-like ampullate silk of *Loxosceles*	–	–	Scopula	–	–	–	Tenent setae of Erythraeidae	–
	Nano-fibrils	Cribellate silk; aciniform silk; particle catching setae	–	–		–	Silk	–	Silk	–

Suction	*Sucker*	–	–	–		–	–	–	Adanal suckers; Gnathosomal sucker plate	–
Adhesive secretion	*Viscid glue*	Aggregate gland secretion; Scytodidae spit; pyriform gland secretion in Pholcidae	–	–	Glandular setae; glandular pedipalps	–	Brood sac	–	Cunaxidae capture silk (?); glandular setae in Epicriidae	–
	Solidifying glue	Pyriform gland secretion([b]); Cylindrical gland secretion; Mygalomorph silk; Mating plug	Egg sac; Spermatophore pad[b]	Egg sac; Spermatophore pad[b]	Body secretion for soil crypsis	Spermatophore pad[b]; Mating plug	Silk; spermatophore pad[b]	?	Pedicels in Uropodina deutonymphs[b]; cement cone in ticks; silk; spermatophore pad[b]; body secretion for soil crypsis	?

Please remark that the particular devices may be restricted to some lineage within the clade (for details see main text). In Palpigradi no attachment devices are known besides the pincer-like chelicerae, claws and empodium, hence this order is not listed here

[a]May be supported by fluid secretions

[b]Applied in a mushroom-like structure

ons, camel spiders, ticks (Ixodida and Holothyrida), mites (Opilioacarida, Mesostigmata, Sarcoptiformes and Trombidiformes), harvestmen nymphs, prenymphs of scorpions, whip spiders and whip scorpions, and, very rarely, in spiders. Hairy adhesive pads are found in spiders, harvestmen, mites (Trombidiformes), and hooded tick spiders. A unique micro-patterned adhesive pad ('smooth' pad with spatulae) occurs in whip spiders. Suckers are only present in mites (Sarcoptiformes and Trombidiformes). Solidifying glue is produced by spiders, whip spiders, whip scorpions, scorpions, harvestmen, pseudoscorpions, ticks (Ixodida), and mites (Mesostigmata, Sarcoptiformes, and Trombidiformes). Viscid glue is produced by spiders, harvestmen, and mites (Mesostigmata and Trombidiformes).

Figure 1.2 presents the distribution of adhesive foot pads in arachnids. In every arachnid order, except Palpigradi, adhesive foot pads have evolved, either smooth or hairy. Due to the uncertainty of the phylogenetic relationships (see conflicting hypotheses and discussion in Shultz 2007; Regier et al. 2010; Pepato et al. 2010; Garwood and Dunlop 2014; Sharma et al. 2014), it is hard to evaluate which of these structures are homologous, in most cases. The relationships among the Tetrapulmonata (Uropygi, Amblypygi + Araneae) and Parasitiformes are well supported and a homology of the arolia within these clades is likely. Recently, the close relationship of scorpions with Tetrapulmonata (=Arachnopulmonata) has been supported by both morphological and molecular based studies (Regier et al. 2010; Sharma and Wheeler 2014; Sharma et al. 2014; Klußmann-Fricke and Wirkner 2016). Therefore, we suggest the pulli-carrying behavior and the correlated occurrence of smooth adhesive pads in the prenymphal instar as synapomorphic for this clade. The hairy pads of spiders, harvestmen and Trombidiformes have multiple origins within the orders, as well as the smooth pads of Acariformes.

In the following we give some more detail on the distribution of the attachment devices and glues within the three most diverse arachnid groups.

1.5.1 Spiders

In spiders there is a multiple origin of structural adhesives (Fig. 1.3). There is multiple evolution of particle catching microtrichia that serve camouflage (Duncan et al. 2007). The occurrence of hairy adhesive pads is strongly correlated with the evolution of a free hunting life style (Wolff et al. 2013). Nonetheless the fine structure of these is strikingly similar. In the most diverse clade of web-building spiders, the Orbiculariae, viscid glue has evolved, which is regarded as a key innovation leading to web shapes of much higher economy and efficiency (Coddington and Levi 1991). Independently, viscid glue has also been evolved in the Pholcidae, but not correlated with an obvious modification in web shapes. This means that in both

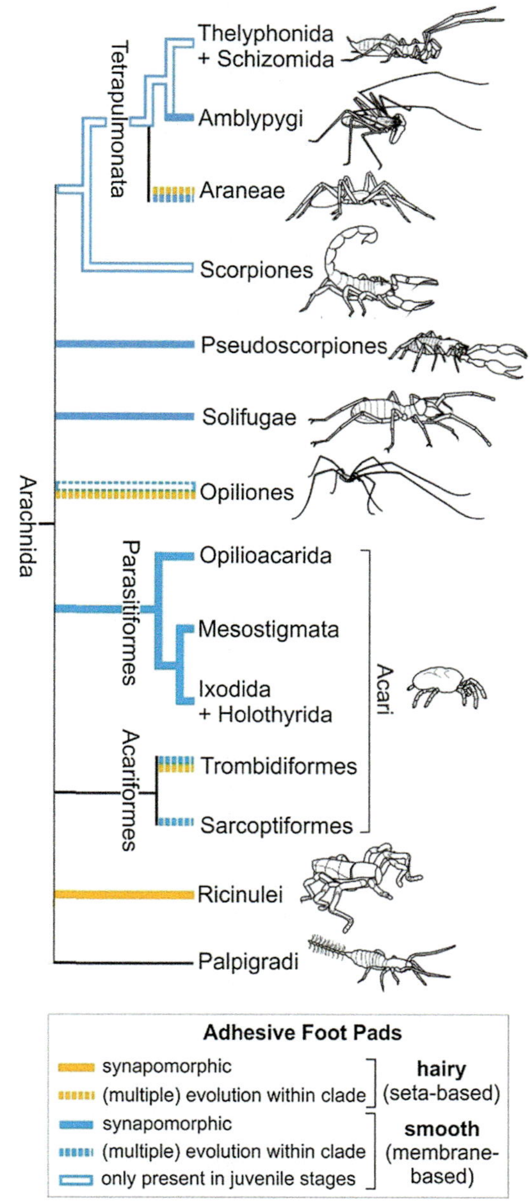

Fig. 1.2 Phylogenetic distribution of adhesive foot pads in Arachnida. Characters are mapped on a compiled tree based on Shultz (2007), Pepato et al. (2010), Regier et al. (2010), Garwood and Dunlop (2014), Sharma and Wheeler (2014). Please note, that the smooth pads of nymphal and adult Amblypygi are functional divergent ('patterned pad', see Chap. 3)

free hunting and web building spiders, many species rely on adhesion to subdue their prey, but with a totally different mechanism. Both free hunting and web-building spiders utilize hardening glue to attach silk threads, egg sacs or seal female genitalia to secure paternity. Another very important glue is the composite piriform gland silk, which is used to anchor silk threads.

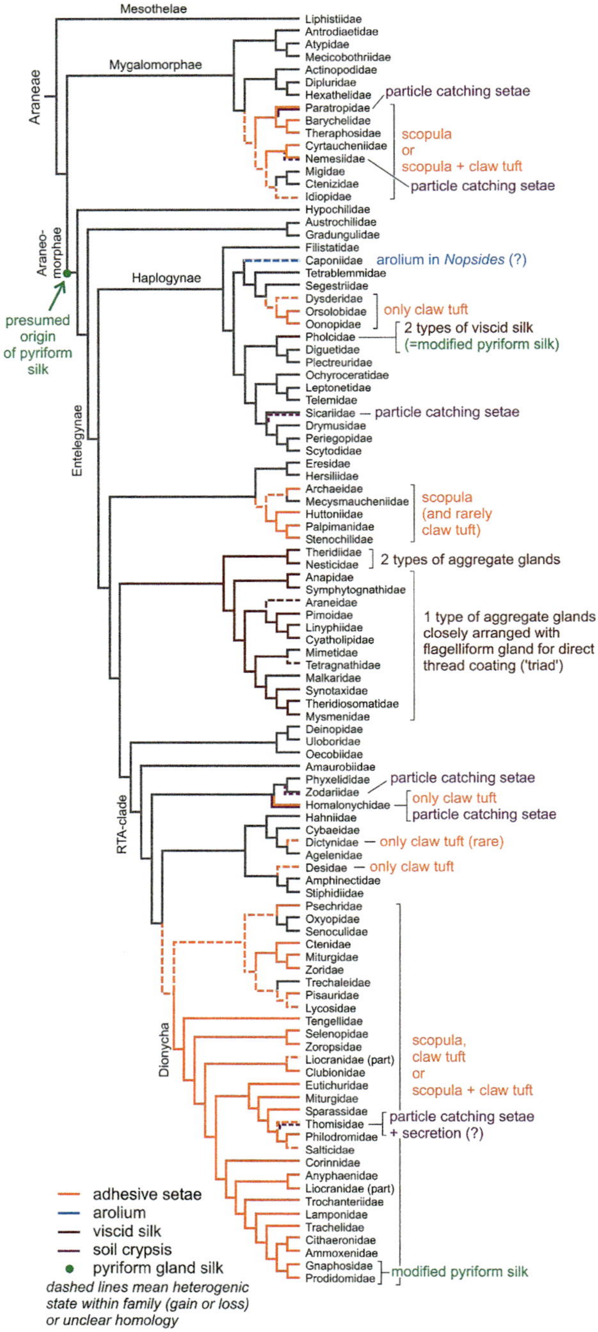

Fig. 1.3 Phylogenetic distribution of adhesive structures and secretions in spiders (Araneae). Compiled cladogram based on Coddington and Levi (1991), Griswold (1993), Coddington (2005), Ramírez (2014), Bond et al. (2014)

1.5.2 *Harvestmen*

Harvestmen evolved both hairy and smooth adhesive pads, and viscid glue (Fig. 1.4). The hairy pads (scopulae) are restricted to some families/sub-families of the Laniatores. Phylogenetic character mapping indicates that they are not homologous (except those of Biantidae and Samoidae). An arolium (smooth adhesive pad) is present in the hind legs of nymphs of Grassatores, and has only been reduced in Sandokanidae. An arolium is also present in Cladonychidae, but presumably it is independently evolved. The nymphal arolium does not play a role during locomotion, but is used to attach to overhanging surfaces prior to molting (Gnaspini 2007).

Viscid glue extruded from pedipalpal glands is used as a means of prey capture in the Palpatores-clade (Wolff et al. 2016). Its use is correlated with modifications of the pedipalp and is of the most significant importance in many Dyspnoi and Neopilionidae. However, some families or species have totally lost this character due to specialization on slow moving prey or dead matter. Pedipalpal glands extruding viscid glue have independently been evolved in nymphs of some Gonyleptidae and Cosmetidae, presumably also to serve prey capture. In contrast to Palpatores, in which the glue is extruded from specialized setae, the glue is emerging from gland openings directly on the pedipalp cuticle.

1.5.3 *Mites and Ticks*

Below the gross phylogenetic relationships of mites and ticks, as well as the distribution of attachment pads and adhesive secretions within this group, is shown (Fig. 1.5). Due to the enormous diversity of this group, it is not within the scope of this book to give a comprehensive overview about all of such structures, and it is impossible to clarify overall homologies. For instance it is unclear, whether the tenent setae occurring in many Trombidiformes are homologous, but it is clear that the smooth adhesive footpads have evolved for multiple times among both Trombidiformes and Sarcoptiformes. Homology of the arolia among Parasitiformes might be probable.

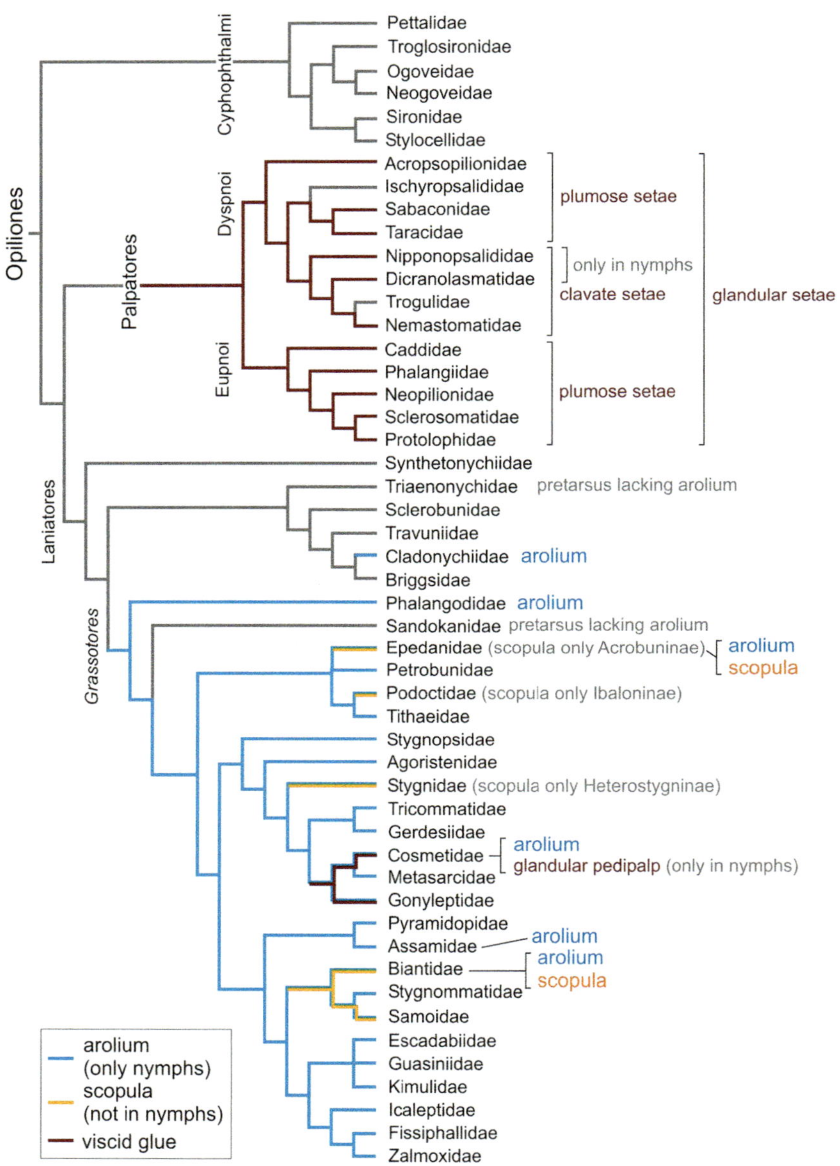

Fig. 1.4 Phylogenetic distribution of attachment devices and adhesive secretions in harvestmen (Opiliones). Compiled cladogram based on Shultz and Regier (2001), Sharma and Giribet (2011), Hedin et al. (2012), Schönhofer (2013), Groh and Giribet (2014), Giribet and Sharma (2015). Characters may be non-homogeneously distributed within lineages (not displayed)

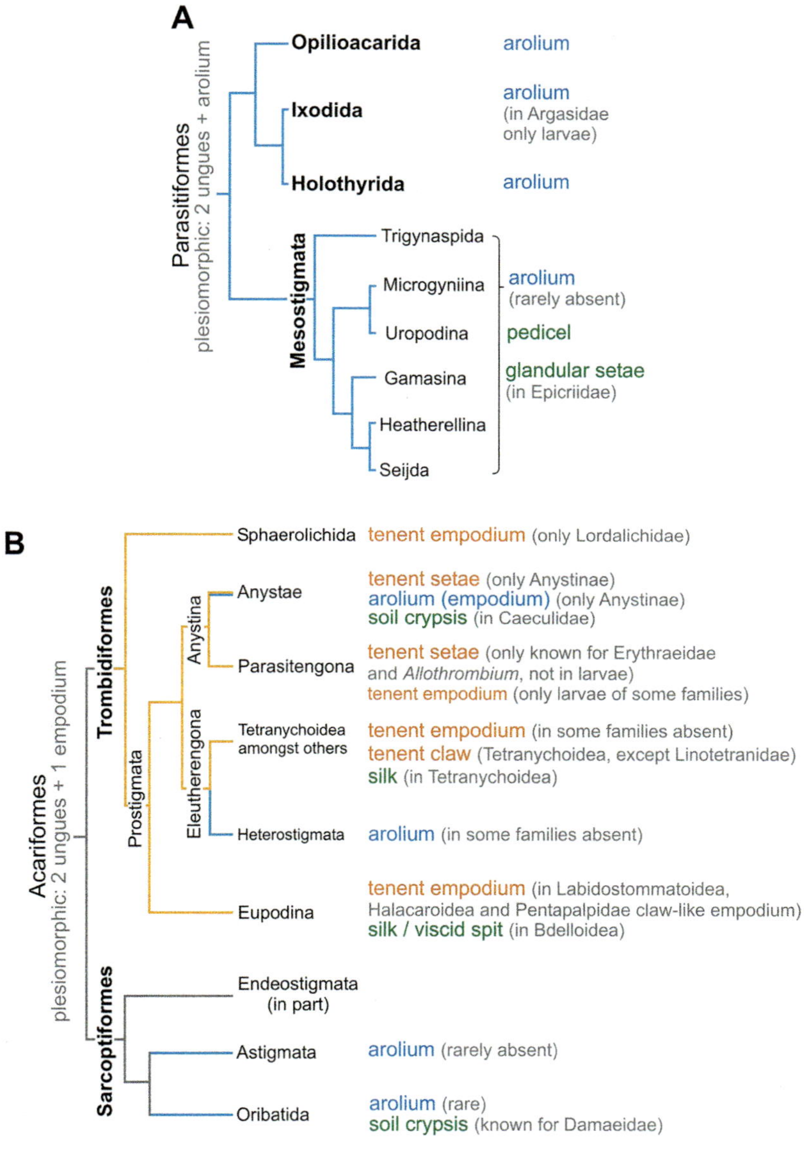

Fig. 1.5 Distribution of adhesive pads in ticks and mites (*orange*: hairy; *blue*: smooth) on the background of phylogenetic relationships of these groups. The mapping marks lineages, in which adhesive pads occur, but they are not necessarily autapomorphic for the particular lineage. Please, note that the marked homology of the structures is a hypothesis, but since some clades are nested within the marked lineages that lack the pads, the organs may also have multiple origin. The occurrence of other noteworthy attachment devices or secretory products is indicated, printed in *green*. (**a**) Parasitiformes, compiled tree is based on Lekveishvili and Klompen (2004), Klompen et al. (2007), Dunlop and Alberti (2008); orders are printed in *bold* font. Opilioacarida, Ixodida and Holothyrida are comparably small orders, and hence not further resolved. (**b**) Acariformes, compiled tree is based on O'Connor (1984), Norton et al. (1993), Dunlop and Alberti (2008), Dabert et al. (2010); orders are printed in *bold* font

References

Betz O, Kölsch G (2004) The role of adhesion in prey capture and predator defence in arthropods. Arthropod Struct Dev 33(1):3–30

Beutel RG, Friedrich F, Yang XK, Ge SQ (2013) Insect morphology and phylogeny: a textbook for students of entomology. Walter de Gruyter, Berlin/Boston

Bird TL, Wharton RA, Prendini L (2015) Cheliceral morphology in Solifugae (Arachnida): primary homology, terminology, and character survey. Bull Am Mus Nat Hist 916(1):1–356

Blick T, Harvey MS (2011) Worldwide catalogues and species numbers of the arachnid orders (Arachnida). Arachnologische Mitteilungen 41:41–43

Bond JE, Garrison NL, Hamilton CA, Godwin RL, Hedin M, Agnarsson I (2014) Phylogenomics resolves a spider backbone phylogeny and rejects a prevailing paradigm for orb web evolution. Curr Biol 24(15):1765–1771

Cloudsley-Thompson JL (1961) Some aspects of the physiology and behaviour of Galeodes arabs. Entomol Exp Appl 4(4):257–263

Coddington JA (2005) Phylogeny and classification of spiders. In: Ubick D, Paquin P, Cushing PE, Roth V (eds) Spiders of North America: an identification manual. American Arachnological Society, Poughkeepsie, pp 18–24

Coddington JA, Levi HW (1991) Systematics and evolution of spiders (Araneae). Annu Rev Ecol Syst 22:565–592

Dabert M, Witalinski W, Kazmierski A, Olszanowski Z, Dabert J (2010) Molecular phylogeny of acariform mites (Acari, Arachnida): strong conflict between phylogenetic signal and long-branch attraction artifacts. Mol Phylogenet Evol 56(1):222–241

Dashman T (1953) Terminology of the pretarsus. Ann Entomol Soc Am 46(1):56–62

De Meijere JC (1901) Über das letzte Glied der Beine bei den Arthropoden. Zool Jahrb Anat 14:417–476

Donovan BJ, Paul F (2005) Pseudoscorpions: the forgotten beneficials inside beehives and their potential for management for control of varroa and other arthropod pests. Bee World 86(4):83–87

Duncan RP, Autumn K, Binford GJ (2007) Convergent setal morphology in sand-covering spiders suggests a design principle for particle capture. Proc R Soc B 274(1629):3049–3057

Dunlop J (2002) Character states and evolution of the chelicerate claws. In: Toft S, Scharff N (eds) European arachnology 2000 (Proceedings of the 19th European Colloquium of Arachnology, Århus 17–22 July 2000), Aarhus University Press, Aarhus, pp 345–354. ISBN 87 7934 001 6

Dunlop J, Alberti G (2008) The affinities of mites and ticks: a review. J Zool Syst Evol Res 46(1):1–18

Fernández R, Giribet G (2015) Unnoticed in the tropics: phylogenomic resolution of the poorly known arachnid order Ricinulei (Arachnida). R Soc Open Sci 2(6):150065

Garwood RJ, Dunlop J (2014) Three-dimensional reconstruction and the phylogeny of extinct chelicerate orders. PeerJ 2:e641

Giribet G, Sharma PP (2015) Evolutionary biology of harvestmen (Arachnida, Opiliones). Annu Rev Entomol 60:157–175

Giribet G, McIntyre E, Christian E, Espinasa L, Ferreira RL, Francke ÓF, Harvey MS, Isaia M, Kováč Ĺ, McCutchen L (2014) The first phylogenetic analysis of Palpigradi (Arachnida)–the most enigmatic arthropod order. Invertebr Syst 28(4):350–360

Gnaspini P (2007) Development. In: Pinto-da-Rocha R, Machado G, Giribet G (eds) Harvestmen: the biology of Opiliones. Harvard University Press, Cambridge, MA, pp 455–472

Gorb S (2001) Attachment devices of insect cuticle. Springer Science & Business Media, Dordrecht/Boston/London, 305 pp

Gorb SN (2011) Biological fibrillar adhesives: functional principles and biomimetic applications. In: da Silva LFM, Öchsner A, Adams RD (eds) Handbook of adhesion technology. Springer, Berlin/Heidelberg, pp 1409–1436

Griswold CE (1993) Investigations into the phylogeny of the lycosoid spiders and their kin (Arachnida: Araneae: Lycosoidea). Smithsonian Contrib Zool 539:1–39

Groh S, Giribet G (2014) Polyphyly of Caddoidea, reinstatement of the family Acropsopilionidae in Dyspnoi, and a revised classification system of Palpatores (Arachnida, Opiliones). Cladistics 31(3):277–290

Harvey MS (2003) Catalogue of the smaller arachnid orders of the world: Amblypygi, Uropygi, Schizomida, Palpigradi, Ricinulei and Solifugae. CSIRO Publishing, Melbourne, 400 pp

Hedin M, Tsurusaki N, Macías-Ordóñez R, Shultz JW (2012) Molecular systematics of sclerosomatid harvestmen (Opiliones, Phalangioidea, Sclerosomatidae): geography is better than taxonomy in predicting phylogeny. Mol Phylogenet Evol 62(1):224–236

Holway RT (1935) Preliminary note on the structure of the pretarsus and its possible phylogenetic significance. Psyche 42(1):1–24

Homann H (1957) Haften Spinnen an Einer Wasserhaut? Naturwissenschaften 44(11):318–319

Kästner A (1941) 2. Ordnung der Arachnida: Pedipalpi Latreille = Geißel-Scorpione. In: Kükenthal W, Krumbach T (eds) Handbuch der Zoologie: Chelicerata, vol 3. Walter de Gruyter & Co., Berlin, pp 1–76

Klompen H, Lekveishvili M, Black WC (2007) Phylogeny of parasitiform mites (Acari) based on rRNA. Mol Phylogenet Evol 43(3):936–951

Klußmann-Fricke BJ, Wirkner CS (2016) Comparative morphology of the hemolymph vascular system in Uropygi and Amblypygi (Arachnida): complex correspondences support Arachnopulmonata. J Morphol 277(8):1084–1103

Kury AB (2015) Classification of Opiliones. Museu Nacional/UFRJ website. Online at: http://www.museunacional.ufrj.br/mndi/Aracnologia/opiliones.html. Accessed 27 Feb 2016

Lekveishvili M, Klompen H (2004) Phylogeny of infraorder Sejina (Acari: Mesostigmata). Zootaxa 629:1–19

Maggenti MA, Maggenti AR, Gardner SL (2005) Online dictionary of invertebrate zoology: complete work. Armand R Maggenti Online Dictionary of Invertebrate Zoology Paper 2. doi:http://digitalcommons.unl.edu/onlinedictinvertzoology/2

Maxwell M (1978) Two rapid and simple methods used for the removal of resins from 1.0 μm thick epoxy sections. J Microsc 112(2):253–255

Nachtigall W (1974) Biological mechanisms of attachment: the comparative morphology and bioengineering of organs for linkage, suction, and adhesion. Springer, Berlin/Heidelberg, 194 pp

Norton RA, Kethley JB, Johnston DE, O'Connor BM (1993) Phylogenetic perspectives on genetic systems and reproductive modes of mites. In: Wrensch DL, Ebbert MA (eds) Evolution and diversity of sex ratio in insects and mites. Chapman & Hall, New York, pp 8–99

OConnor BM (1984) Phylogenetic relationships among higher taxa in the Acariformes, with particular reference to the Astigmata. In: Griffiths DA, Bowman CE (eds) Acarology VI, vol I. Ellis-Horwood Ltd., Chichester, pp 19–27

Pepato AR, da Rocha CE, Dunlop JA (2010) Phylogenetic position of the acariform mites: sensitivity to homology assessment under total evidence. BMC Evol Biol 10(1):235

Ramírez MJ (2014) The morphology and phylogeny of dionychan spiders (Araneae, Araneomorphae). Bull Am Mus Nat Hist 390:1–374

Regier JC, Shultz JW, Zwick A, Hussey A, Ball B, Wetzer R, Martin JW, Cunningham CW (2010) Arthropod relationships revealed by phylogenomic analysis of nuclear protein-coding sequences. Nature 463(7284):1079–1083

Rein JO (2015) The scorpion files. In: Roskov Y, Abucay L, Orrell T et al (eds) Species 2000 & ITIS catalogue of life. Naturalis, Leiden. Digital resource at www.catalogueoflife.org/col

Schönhofer AL (2013) A taxonomic catalogue of the Dyspnoi Hansen and Sørensen, 1904 (Arachnida: Opiliones). Zootaxa 3679(1):1–68

Seiter M, Wolff J, Hörweg C (2015) A new species of the South East Asian genus Sarax Simon, 1892 (Arachnida: Amblypygi: Charinidae) and synonymization of Sarax mediterraneus Delle Cave, 1986. Zootaxa 4012(3):542–552

Selden PA, Shear WA, Sutton MD (2008) Fossil evidence for the origin of spider spinnerets, and a proposed arachnid order. Proc Natl Acad Sci U S A 105(52):20781–20785

Sharma PP, Giribet G (2011) The evolutionary and biogeographic history of the armoured harvestmen–Laniatores phylogeny based on ten molecular markers, with the description of two new families of Opiliones (Arachnida). Invertebr Syst 25(2):106–142

Sharma PP, Wheeler WC (2014) Cross-bracing uncalibrated nodes in molecular dating improves congruence of fossil and molecular age estimates. Front Zool 11:57

Sharma PP, Kaluziak ST, Pérez-Porro AR, González VL, Hormiga G, Wheeler WC, Giribet G (2014) Phylogenomic interrogation of Arachnida reveals systemic conflicts in phylogenetic signal. Mol Biol Evol 31(11):2963–2984

Shear WA, Palmer JM, Coddington JA, Bonamo PM (1989) A Devonian spinneret: early evidence of spiders and silk use. Science 246(4929):479–481

Shultz JW (2007) A phylogenetic analysis of the arachnid orders based on morphological characters. Zool J Linnean Soc Lond 150(2):221–265

Shultz JW, Regier JC (2001) Phylogenetic analysis of Phalangida (Arachnida, Opiliones) using two nuclear protein-encoding genes supports monophyly of Palpatores. J Arachnol 29(2):189–200

Smith AM, Callow JA (2007) Biological adhesives. Springer Science & Business Media, Berlin/Heidelberg/New York, 284 pp

Smrž J, Kováč Ĺ, Mikeš J, Lukešová A (2013) Microwhip scorpions (Palpigradi) feed on heterotrophic cyanobacteria in Slovak caves–a curiosity among Arachnida. Plos One 8(10):e75989

Vitzthum H (1931) 9. Ordnung der Arachnida: Acari = Milben. In: Kükenthal W, Krumbach T (eds) Handbuch der Zoologie: Chelicerata, vol 3, vol 2. de Gruyter & Co., Berlin, 160 pp

von Byern J, Grunwald I (2010) Biological adhesive systems: from nature to technical and medical application. Springer, Vienna/New York, 304 pp

Weygoldt P (1969) Biology of pseudoscorpions. Harvard University Press, Cambridge, MA, 145 pp

Wolff JO, Nentwig W, Gorb SN (2013) The great silk alternative: multiple co-evolution of web loss and sticky hairs in spiders. Plos One 8(5):e62682

Wolff JO, Schönhofer AL, Martens J, Wijnhoven H, Taylor C, Gorb SN (2016) The evolution of pedipalps and glandular hairs as predatory devices in harvestmen (Arachnida, Opiliones). Zool J Linnean Soc. Online pre-pub. doi:10.1111/zoj.12375

World Spider Catalog (2016) World spider catalog. Natural History Museum Bern. Online at http://wsc.nmbe.ch, version 17.0. Accessed 26 Feb 2016

Chapter 2
Mechanical Attachment Devices

Abstract Rigid cuticular structures that generate high friction by mechanical interlocking with substrate protuberances are the most abundant attachment devices in arthropods. Here we review the different types of such structures in arachnids. The claws of appendages are primarily to increase foothold on walking and climbing substrates, but may also be important means of prey capture, which often comes with structural modifications. Claws are often supported by spines and microtrichia, all of which have different tip diameters. Such a set of differently sized interlocking devices may generate friction on a broad range of substrate roughness from very fine to rather coarse. Other examples of interlocking mechanisms are hooks and platelets on different body parts, some types of genital sealing (mating plugs), and silk threads interlocking with protuberances of substrate surfaces. Clamps and pincers are muscle-driven attachment devices, which are abundant means of prey capture and food handling: the chelicerae, the pedipalpal chelae of scorpions and pseudoscorpions, the raptorial legs (spinated legs or pedipalps for prey capture), and the cucullus ('hood') of Ricinulei. Other mechanical attachment devices are expansion anchors that are first pierced into a substrate, such as prey cuticle, plant epidermis or host skin, and interlocked there. They are represented by some specialized mouth parts. Lock-and-key is a system, in which two specific co-adapted structures interlock with each other. They can be found in some copulation organs. Wrapping around the substrate is an effective means of attachment, as it can highly enhance the action of friction. This mechanism is frequently used by long-legged harvestmen, when walking on thin cylindrical surfaces like grass or herb stems. It is further utilized by some spiders which tightly wrap their prey in silk.

Most attachment structures use the principle of mechanical interlocking with substrate structures. This principle is associated with low costs as no material discharge is necessary (as in the case with adhesive secretions) and the structures are relatively simple. Furthermore, these structures or organs often exhibit a high universality and efficacy of usage, especially on rough substrates. This may explain why it is the

Electronic supplementary material The online version of this chapter (doi:10.1007/978-3-319-45713-0_2) contains supplementary material, which is available to authorized users Videos can also be accessed at http://www.springerimages.com/videos/978-3-319-45712-3.

J.O. Wolff, S.N. Gorb, *Attachment Structures and Adhesive Secretions in Arachnids*, Biologically-Inspired Systems 7,
DOI 10.1007/978-3-319-45713-0_2

most widespread principle. Despite of their abundance and importance, these structures are often highly neglected in biomechanic studies.

2.1 Claws

Claws are the most ubiquitous attachment devices, and equally found in arthropods, vertebrates and some annelids. The shape and material of the claws may vary, but the basic structure is strikingly similar. The base is broad and flexibly hinged to absorb stress and provide certain adaptability and resistance of the attachment during movement. The shaft is curved ventrally, which enhances interlocking capabilities by exposing the tip to cavities on corrugated surfaces. The tip is highly pointed. Its diameter is crucial for the interlocking ability on a broad range of substrates and should be preferably small (Dai et al. 2002; Ditsche-Kuru et al. 2012; Labonte and Federle 2015). The efficacy is dependent on the resistance of the structure, because too strong deformation under load would lead to a loss of the interlocking mechanism. Hence, the material must be very stiff at the tip the sharper it is, in order to prevent deformation. However, a rigid claw exhibits a high stress concentration in the tip, which increases the risk of breakage the more pointed the tip is (Labonte and Federle 2015). To minimize this risk, the material strength of the claw tip is usually enhanced e.g. by the incorporation of metal ions (Fontaine et al. 1991; Schofield et al. 2003) or by specific material structure (Bar-On et al. 2014). Additionally, there is often an elastic hinge at the claw articulation or an underlying elastic material layer that works as a stress absorber. In some cases, the claw may additionally be sub-segmented, as it is the case in solifuges (see below), which is presumably a further mean to reduce stress concentration at the claw tip.

Claws are present in all arachnid orders and have only got lost in some specialized parasitic mites. Furthermore, they are lacking in the prenymphal stages of some groups (Wolff et al. 2015a). It is presumed that the ancestral state in chelicerates is a foot with two large, hinged claws and a smaller claw-like empodium that is fused to the pretarsus (Dunlop 2002) as it is the case in scorpions (Fig. 2.1a).

In spiders, this state is found in the four pairs of walking legs (Fig. 2.1d; Videos 2.1 and 2.2), however, in most spiders that exhibit hairy adhesive foot pads (clawtufts) the empodium is highly or totally reduced (Fig. 2.1e). Both claws and empodium can be smooth or pectinate (serrated). In orb-webspiders and their relatives (Araneidae, Theridiidae, Linyphiidae), the empodium is elongated and forms a carabiner mechanism together with serrated bristles (Foelix 1970) (see Sect. 2.2 and Fig. 2.3). The two main claws are often slightly dissimilar. For instance, in jumpingspiders (Salticidae) the pro-lateral claw bears a higher number of teeth with finer interspaces, and may be used to scale silk threads (Hill 1977, 1978). In some spiders, the prolateral claw (proclaw) is extremely enlarged and hook-like, a modification which is related to a specific prey capture behavior. In the Gradungulidae and Trogloraptoridae, the pro-claw can be highly flexed against the tarsus, hence forming a kind of clamp (Forster et al. 1987; Griswold et al. 2012). *Progradungula carraiensis* (Gradungulidae) uses the claws to pick up running prey from the ground, which is then thrown into a small cribellate capture web, held pre-stressed with the hind legs (Gray 1983). In *Trogloraptor,* the tarsi exhibit a unique sub-segmentation, such that the distal tips can be flexed with the claw forming a presumed raptorial prey capture apparatus (Griswold et al. 2012). Also in *Doryonychus raptor* (Tetragnathidae), the pro-claw of the front legs is extremely enlarged (Fig. 2.1f) and used as a hook to catch passing flying prey (Gillespie 1991). This special modification comes with the total loss of the ability to build webs and produce sticky silk (Gillespie 1991). In the crab spider *Heterogriffus berlandi* (Thomisidae), a comparable claw dissimilarity can be found in the front legs (Platnick 1976), which is probably related to a prey capture function, too, since crab spiders are ambushing hunters grasping passing prey with their massive front legs.

In certain spiders, claws may also be present in the pedipalps, especially in the Mygalomorphae, but (reduced) also in some Araneomorphae.

In whip-spiders the walking legs (legs II–IV) are equipped with two claws (Fig. 2.1c; Video 2.3) and, in the basal lineages, with a median attachment pad (see Chap. 7) (Dunlop 2002). The sharpness of the claw seems to differ between amblypygids that have adhesive pads and those that lack it. Further, in the whip-spiders with the median attachment pad, the claws highly rotate during protraction, such that they eventually face each other (Wolff et al. 2015b) (Video 7.9), a mechanism which can also be observed in some other arachnids and insects. This is presumably primarily to give room for the attachment pad, but the claws also may act as a clamp in these cases. Whip-scorpions show the ancestral state with two main claws and the hook-like empodium (Fig. 2.1b; Video 2.4). The first leg pair of whip-spiders and whip-scorpions lacks the claw, since it is highly modified into antenna-like sensory legs.

Among harvestmen, species of the Cyphophthalmi, Eupnoi, and Dyspnoi bear only a single claw in their pedipalps and walking legs, which is immovably fused with the pretarsus. Most Laniatores bear two claws in the two posterior leg pairs (Fig. 2.1m), which are the primary (and often exclusive) locomotory appendages in this sub-order (Pinto-da-Rocha et al. 2007), and a single claw in the front legs (Fig. 2.1n) and pedipalps. In some species also an empodium is present (Fig. 2.1l). The Triaenonychidae and Travuniidae (Laniatores) only bear a single claw in the hind legs, which may bear lateral denticles (Fig. 2.1o) or be highly modified to a furcated or spoon like structure (Roewer 1923, 1935). Strong modifications of the claws are also present in the hind legs of nymphs of Cladonychiidae with a multiplication of the claws and an arolium-like protuberance of the pretarsus (see Chap. 7). In adult Cladonychiidae the hind leg claws are tilted towards each other (Fig. 2.1m), the function of that is dubious, since the claws cannot be moved against each other. The claws of the pedipalps are highly elongated in most Laniatores and can be highly flexed against the tarsus, which is an important means of prey capture (raptorial claw, see Fig. 2.8c, d). In contrast, the pedipalpal claws are partly or totally reduced in most Dyspnoi and some Neopilionidae, in which the pedipalps are modified into sticky tentacles for prey capture (see Chap. 8).

In solifuges, the paired claws are very long and uniquely sub-segmented in their apical part (Fig. 2.1h) (Kästner 1933). This flexible hinge of the claw tip may have

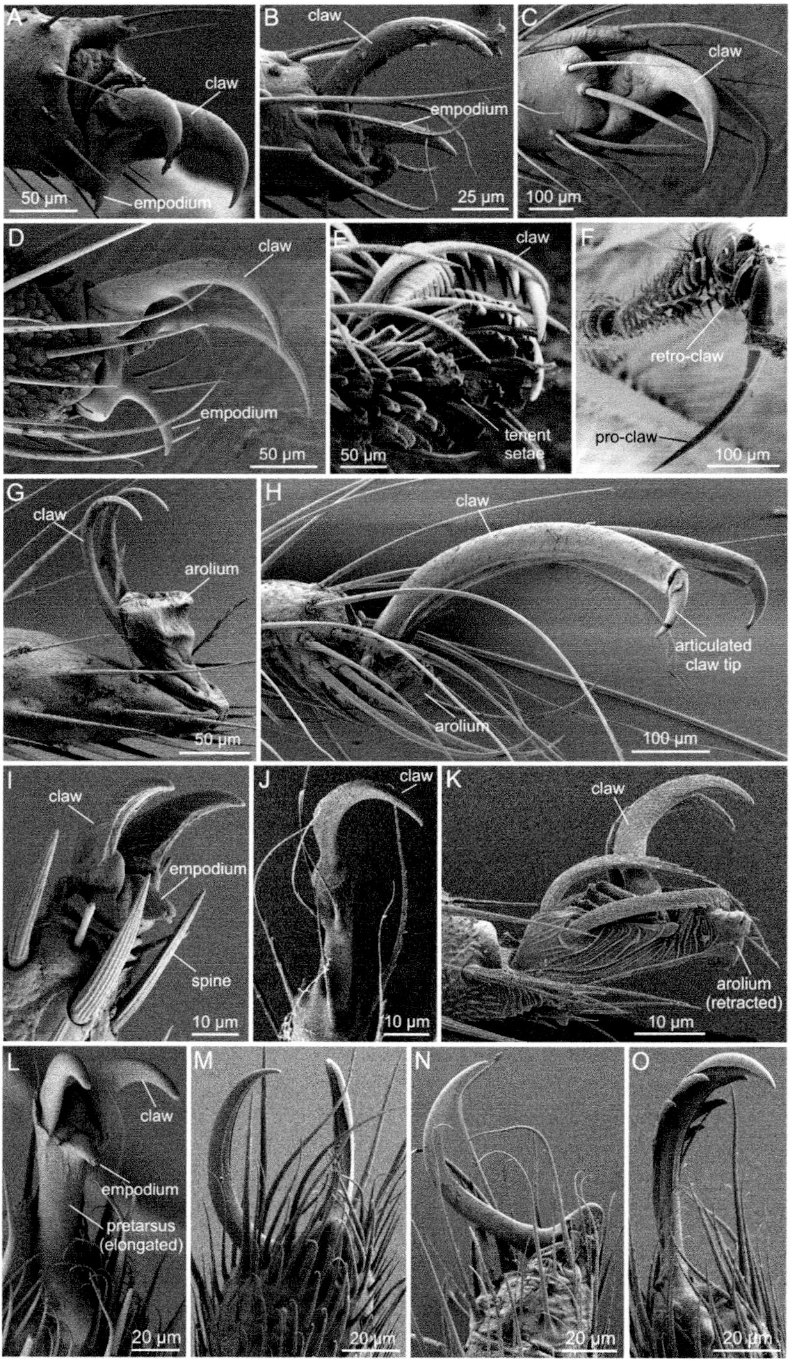

Fig. 2.1 Claws of walking legs in arachnids. (**a**, **b**) In its plesiomorphic (basal) state the arachnid pretarsus bears two claws and a median hook-like empodium, as in scorpions (**a**, here protonymph of *Liocheles australasiae*) or whip-scorpions (**b**, here Schizomida, unknown species). (**c**) Whip-spiders have lost the empodium in favor of an arolium, which, however, is secondarily reduced in most species, as in this *Phrynus longipes*. (**d**–**f**) Claws in spiders are diverse: (**d**) the basal Mesothelae (here: *Vinathela tonkinensis*) still show the plesiomorphic state, however, with the main claws carrying additional teeth (see also Video 2.1); (**e**) the Orsolobidae (here: *Falklandia rumbolli*)

the function, to reduce the stress acting on the tip, and in turn to prevent claw breakage and material wear. The claws are smooth, lack any denticles and are highly curved. In contrast to legs II–IV, the first leg pair of solifuges is rather thin and weak and in many species not used for locomotion, but rather for sensing as in whip-scorpions. Accordingly, the pretarsus and claws are often reduced or totally lacking in the first leg pair; only in the Rhagodidae they are still fully developed (Kästner 1933; Klann 2009).

Various states of the pretarsal claws occur among mites: three, two, one or no claws. The three-clawed state (two claws + empodium) is assumed to be plesiomorphic for the Acariformes (Fig. 2.1i), whereas it is the two-clawed state in the Parasitiformes (Fig. 2.1k) (Dunlop 2002). The claws and/or the empodium may also be modified in an adhesive apparatus carrying spatulate microtrichia, as in the Tetranychidae (see Chap. 3). In Anystinae (Trombidiformes: Anystidae), the empodium is modified into an arolium (see Chap. 7), whereas in the other anystids it is claw-like (Meyer and Ueckermann 1987). In the related Parasitengona-clade, the empodium is only present in larvae and may represent an adhesive pad (at least in Erythraeidae, see Chap. 3) or a claw (presumably secondary). The claws are usually smooth in the orders of the Parasitiformes clade and in the Sarcoptiformes, but may be highly pectinate in the Trombidiformes. In aquatic mites (Trombidiformes: Hydracarina and Halacaroidea) claws are the exclusive foot attachment organs and often enlarged and equipped with denticles (Vitzthum 1943; Alberti and Coons 1999). It was proposed that the claw structure may reflect the ecological niche, with the claws being longer in mites of the intertidal zone than in those of the supralittoral (Pugh et al. 1987). In the large soil mite sub-order Oribatida (Sarcoptiformes), adhesive pads are very rare and the claws are the dominant means of attachment (Fig. 2.1j). Due to the small tip diameter, they can generate high friction even on surfaces with a very small roughness, where the claws of larger arthropods fail (Heethoff and Koerner 2007). In the oribatid *Archegozetes longisetosus*

Fig. 2.1 (continued) exhibit spoon-like claws with multiple rows of differently sized teeth on the lateral edges, the empodium is lacking in favor of the tenent setae (image: W. Nentwig, with kind permission); (**f**) strongly modified raptorial pro-lateral claw of *Doryonychus raptor*, note the strongly reduced retro-lateral claw (image: R. Gillespie, with kind permission, adapted from Gillespie 1991). (**g**) Pseudoscorpions (here: *Neobisium spelaeum*) have lost the empodium in favor of the large arolium, the claws can be highly retracted due to the slanted shape of the distal tarsus. (**h**) Solifugae (here: unidentified species of Solpugidae) bear an arolium instead of a sclerotized empodium, too; the paired claws are enormously enlarged and uniquely exhibit articulated tips, presumably an adaptation to fast climbing through a rocky terrain to reduce wear of the claw tip. (**i**) Among Trombidiformes, the plesiomorphic foot configuration can be found in the basal Caeculidae (here: *Caeculus* sp.). (**j**) The mites of the order Sarcoptiformes bear 0–3 claws, this unidentified Oribatida exhibits a single claw on each leg. (**k**) Among the Parasitiformes the plesiomorphic state is two claws and a median arolium, as in the basal Opilioacarida (here: *Opilioacarus texanus*). (**l–o**) Harvestmen show a diversity of claw configurations: (**l**) some Laniatores bear two claws and a median empodium in the two posterior leg pairs, as species of the family Sandokanidae (here: leg IV of a nymph of *Gnomulus* sp.); (**m, n**) in the Cladonychiidae (here: *Holoscotolemon querilhaci*) the hind legs carry two claws, which are strangely tilted towards each other (see (**m**)), whereas the front legs exhibit a single claw that is immovably fused with the pretarsus (see (**n**)) as typical for Laniatores front legs or all legs of Eupnoi and Dyspnoi; (**o**) In the Triaenonychidae the hind legs bear only a single claw, but which is equipped with denticles on the lateral edges (here: leg IV of a nymph of *Paranonychus* sp.)

(Trhypochthoniidae) the claw protractor muscle is very strong (about 1200 kN/m^2), and the attachment apparatus can withstand loads over 1100 times of the mite body weight without damage (Heethoff and Koerner 2007). Presumably, the cuticle of such claws is reinforced by the inclusion of metal ions, but this has not been studied in mites so far.

This overview illustrates that claws may vary considerably in their relative size, curvature, sharpness, denticulation and, presumably, hardness, all of which may be due to adaptive effects. Further broad comparative approaches are needed to understand the evolutionary factors that shape claws, not at least, to find optimal shapes for biomimetic hooks, tailored for specific applications.

2.2 Spines and Pointed Hairs

2.2.1 Spines

Spines (stiff, pointed cuticular protuberances) are frequently found and may act similarly to claws, but usually on smaller substrate structures, because of a smaller tip diameter. Hence they might very effectively support the claws, in order to generate high friction on rough surfaces. Arrays of spines may also improve the foothold in a complex microenvironment (Spagna et al. 2007) (i.e. herbal layer with cylindrical plant stems in short distances and different angles), or as a prey capture basket (building a barrier for struggling prey).

Spines for Retaining Prey In spiders, the spines of the walking legs usually lay flat (Fig. 2.6b), but get erected during attack situations (Rovner 1980; Eggs et al. 2015) (Video 2.5). They assist prey capture by building a switchable physical barrier in the leg capture basket (Fig. 2.6a–d). Since the long erected spines could hinder the locomotion in a structured terrain, such switchability might be very beneficial. The mechanism is based on the hydraulic system of arachnids, which is also used in locomotion. Interestingly, the spines also erect during jumps (Ott and Wolff, unpublished data), which indicates that the erection cannot specifically be controlled. This means, when the hemolymph pressure in the leg exceeds a certain threshold, the spines get erected, which can easily be tested by squeezing a freshly ablated spider leg. Such erectable spines are especially well expressed in hunting spiders, and are usually more numerous and longer in the front legs (Eggs et al. 2015). An exception is represented by lynx spiders (Oxyopidae), which exhibit long spines that are durably erected and immobile (Video 2.6). Probably, their spines are too long and heavy to get erected by a slight increase in the hemolymph pressure, and therefore they have lost their mobility. Since these spiders usually hunt on herbaceous vegetation, the spines may not hinder the leg movements. Long spines are also often present in the pedipalps of solifuges, in which they may serve as a prey capture basket, too. Capture baskets built by spinated legs, analogous to those of spiders, are known from some predatory insects, like dragonflies (Gorb 1999), but in insects, the spines are always erected.

Specialized raptorial pedipalps or legs independently evolved in the Pedipalpi (Uropygi+Amblypygi) (Fig. 2.8a, b), harvestmen of the suborder Laniatores (Fig. 2.8c, d) and some Palpatores, and some mites, like the Caeculidae (Trombidiformes) (Fig. 2.8e). However, these spines are not always modified setae. In the Pedipalpi, the spines are cuticular outgrowths and, in harvestmen, their largest proportion is built by the socket, which is highly elevated. A strikingly similar feature of all these structures is that they are distally tilted. At first glance, this seems to be counter intuitive, because a proximal tilting would represent a barbed leg, whose grip can hardly be escaped. Presumably, the release of a struggling prey item from this grip would be more difficult, which would be critical, if the prey turns out to be harmful. Hence, for a predator it is important to control the attachment to the prey, in order to reduce the risk of getting injured.

Spines for Locomotion Other, shorter spines that are located at the ventral side of tarsi of the walking leg are usually immobile. Such spines are present in many arachnids that live on the ground, on rock surfaces or tree trunks, especially in scorpions, whip-scorpions (Fig. 2.2), whip-spiders, some spiders, and some harvestmen (Fig. 2.10b–d). They are often more expressed in the hind legs and may generate friction on rough substrates during forward thrust, especially when the claws are already elevated (the claws usually get elevated before leg movement). In whip-spiders and whip-scorpions, basally thick, pointed spines are standing in two lateral rows on the ventral sides of the tarsal segments. The spines are directed distally which especially supports generation of high friction and forward thrust in the hind legs during locomotion and resting on vertical rough surfaces (Wolff et al. 2015b) (Fig. 2.2). Furthermore, this orientation of the spines is contrary to that of the claws, hence strong attachment can be achieved by counter-acting friction, when the claws are lowered (Fig. 2.2d). Since the claws are usually still elevated, when the foot touches the substrate, the spines are also important to establish foothold (Fig. 2.2b; Video 2.4).

Also in solifuges the ventral sides of walking leg tarsi are equipped with strong spines that play an important role in interlocking and friction generation during locomotion (Kästner 1933). The tarsus of the walking leg can be bent towards the substrate due to its sub-segmentation, which enhances the action of the spines. Some spines may be joined with the underlying cuticle through articulated sockets. This articulation provides an ability to change the spine angle by the change of internal hemolymph pressure. Since most solifuges are epigaeic, and live in stony desert habitats, the spines may represent attachment devices sufficient for the proper mechanical contact with the typically rough substrates. Despite of their often large size, solifuges are able to run very fast (Punzo 1998a), which underlines the importance of a proper grip on the ground substrate.

In orb web-spiders, bent pectinate bristles, also called *'auxiliary claws'*, are present on the distal tarsus underneath the claws (Fig. 2.3c). In interaction with the elongated empodium these work like a carabiner, used to grasp silk threads (Foelix 1970) (Fig. 2.3d).

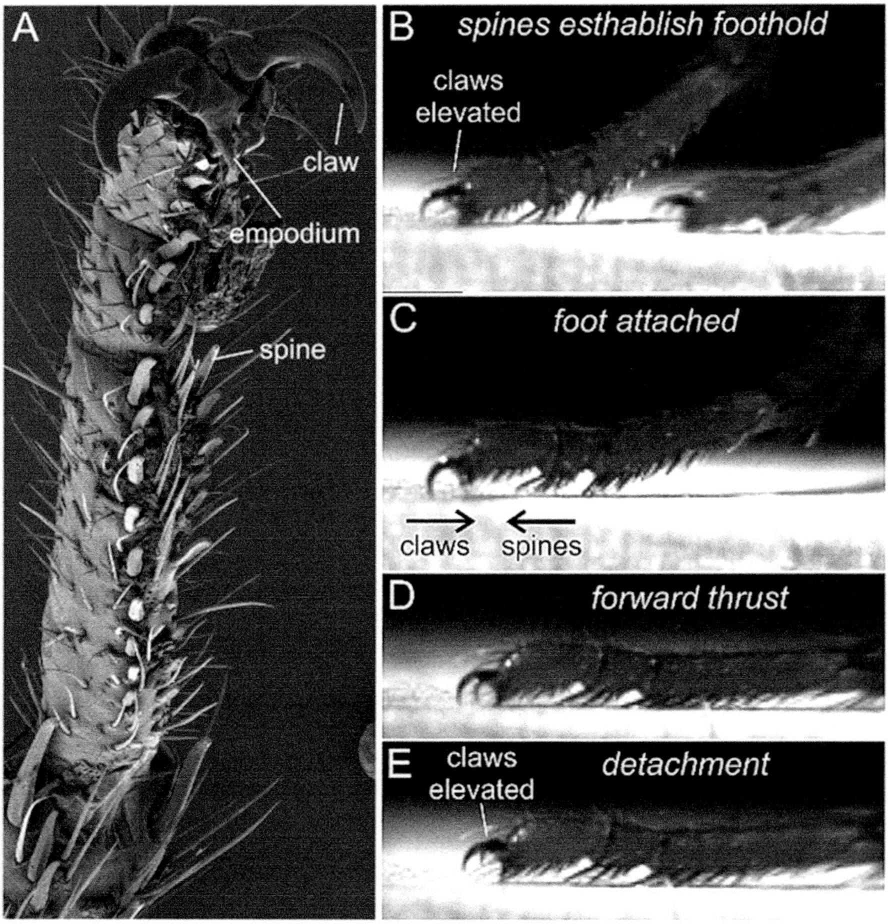

Fig. 2.2 Tarsal spines in whip-scorpions. (**a**) The ventral sides of walking leg tarsi are equipped with strong spines in some arachnids as it is seen in this SEM image of the hind leg tarsus of the whip-scorpion *Typopeltis crucifer*. (**b–e**) Frames of a high speed video sequence of a hind leg of a *T. crucifer,* walking on a wooden bar. The spines act complementary to the claws, to establish foothold and generate high friction during forward thrust. Note the contrary orientation of claws and spines, leading to strong attachment when the claws are lowered. Also note that claws and spines detach at different time points (See Video 2.4)

2.2.2 Microtrichious Setae

In many sheet-web building and hunting spiders, dense brushes of microstructured setae are present on the distal tarsus (Fig. 2.3a). They are flattened and densely covered with tapered microtrichia on the substrate facing side (Fig. 2.3b). These may produce high friction forces on micro rough surfaces due to the interlocking with micro asperities. These 'frictional setae' may thus support the claws and spines, and, if present, tenent setae, to generate friction on a broad range of surface

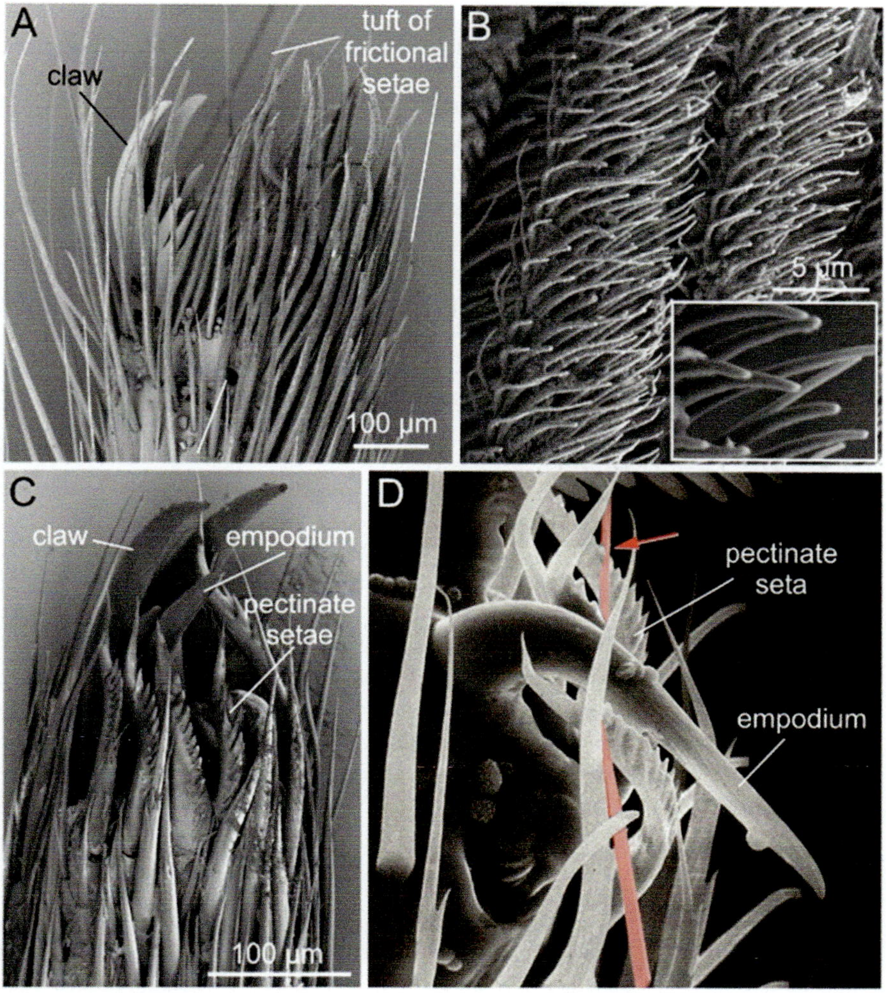

Fig. 2.3 Specialized setae in spiders for interlocking with silk threads. (**a**, **b**) In many spiders, building primitive sheet or funnel webs, like Agelenidae (here: *Malthonica ferruginea*), the distal ventral side of the tarsi is covered with anisotropic microtrichious setae. These facilitate quick movements on the web and may also generate friction on micro rough substrate surfaces. (**c, d**) Spiders building more sophisticated webs, like orb-web spiders (Araneidae, here: *Araneus* spp.) bear massive pectinate spines, which work together with the enlarged, hook-like empodium to grasp silk threads (thread colored *red* in (**d**)); image (**d**) by R. Foelix, with kind permission

roughness scales. These setal brushes also increase the spider maneuverability during the locomotion on sheet webs (Foelix 2011). Presumably, this might be the initial function of these setae (Wolff et al. 2013). They might be a precursor structure from which adhesive setae with spatulate microtrichia (see Chap. 3) have been later evolved (Wolff et al. 2013).

Special microtrichious setae with knobbed tips are present on the dorsal opisthosoma of female wolf spiders (Lycosidae) assisting the attachment of juveniles, which interlock with these setae using their claws and thin silk threads (Rovner et al. 1973). If these setae are removed, the juveniles cannot hold themselves on the mother anymore, which highly affects their survival (Rovner et al. 1973).

Some desert dwelling huntsman spiders (Sparassidae) bear long, stiff setae with short hook-like microtrichia on their pedipalps and chelicerae that together form a basket-like structure used to transport sand during the building of burrows (Foelix et al. 2016). The microtrichia presumably enhance the attachment of sand particles by interlocking with minute surface irregularities.

2.3 Other Interlocking Devices

2.3.1 Hooks and Platelets

Attachment of Soil Particles for Camouflage Some spiders cover themselves by soil particles in order to blend with their environment for camouflage. In spiders of the genus *Cryptothele* (Zodariidae), soil particles are mechanically held in place by curved setae covered by long barbs (Ramírez et al. 2014). In some crab spiders (Thomisidae), environmental particles, detritus and growing fungi cover the body, attached by barbed club-like setae and papillate microtrichia; an adhesive secretion may also be involved (Ramírez 2014; Gawryszewski 2014). Soil crypsis is also known from some burrowing mygalomorphs (Paratropidae, some Nemesiidae). The soil is encrusted at scale-like setae (Raven 1985). Barbs are also present in some harvestmen, like Dicranolasmatidae and Trogulidae, and act complementary to a solidifying secretion to encrust soil particles (see Chap. 8).

In spider species of the genera *Sicarius* (Sicariidae) and *Homalonychus* (Homalonychiidae), a different mechanism of soil crypsis has evolved: here sand grains adhere to very fine fibril-like microtrichia presumably by the action of van-der-Waals forces (see Chap. 4). Soil crypsis is also quite common in harvestmen and some mites. These arachnids utilize a solidifying secretion for this purpose (see Chap. 8). Thus, despite of similar biological functions of different soil particle holding structures, their adhesion mechanisms differ widely.

Barbs for Predator Deterrence Many New World tarantulas (Theraphosidae) possess specialized setae that are discharged in order to deter predators by causing irritations. These, so-called, urticating setae exhibit barbs that are assumed to attach to the skin or cuticle of predators (Bertani and Guadanucci 2013), however, the exact mechanism remains to be studied.

Friction Generation During Peristaltic Locomotion The cuticle of the worm-like mites of the family Nematalycidae (Endeostigmata) bears platelet-like microstructures. These mites live in the interstitial of marine sediments, and the microstructures are assumed to support the peristaltic movements through the spaces between sand grains by interlocking with surface irregularities of the grains in the contraction phase (Bolton et al. 2015).

Attachment to Female Genital Tract In spiders, direct sperm transfer is performed by the modified male pedipalpal tarsus. This bears complex structures that are assumed to specifically interlock with parts of the female genital tract. Harvestmen have a penis structure. In several lineages, its tip (the glans) is equipped with barbs or denticulate processes that may be spread by an increase of hemolymph pressure and retracted by muscles (Pinto-da-Rocha et al. 2007; Schönhofer 2008). These presumably serve to anchor the penis with the female genitalia, to prevent an interruption of the sperm transfer (see Chap. 9). In *Sclerobunus* sp. (Travunioidea), it was observed that the attachment of this male structure is so strong that it is not possible to separate a copulating male from the female without injury (A.L. Schönhofer, pers. communication).

Some arachnids perform an indirect sperm transfer via spermatophores. These are complex secretion products that are formed and deposited in a long process, and include tanned (hardened) structures that may act as interlocking devices to get attached to the genital tract of the female. For instance, in scorpions, the spermatophores bear apically projecting lobes that interlock with the female genital tract, when the female is guided over the spermatophore by the male (Jacob et al. 2004). This is less a lock-and-key mechanism rather than an unspecific mechanical fixation which reduces the ability of the female to interrupt the sperm transfer (Peretti 2004). Also the spermatophore of some pseudoscorpions apically bears a complex sperm capsule with wing- and horn-like protuberances (Kew 1930; Legg 1973; Weygoldt 1969). The protuberances do not seem to play a role in a lock-and-key mechanism, as only the insemination duct enters the female genital opening, however, some kind of interlocking takes place (Weygoldt 1966). In whip-spiders, the spermatophores apically bear horn-, wing- or hook-like protuberances, too (see Fig. 5.3 in Chap. 5). Here, these are primarily assumed to help the female to orientate and find the sperm masses (Weygoldt 2000). However, they may also interlock with the female's genital tract, holding it in place, while the sperm is ripped out (see below). In the large whip-scorpions (Thelyphonida), the spermatophores are equally elaborate as in whip spiders (Weygoldt 1988; Weygoldt and Huber 2013), but here the exact function of hook-like protuberances remains unstudied.

Gripping of Sperm Female amblypygids possess soft eversible appendages (so-called *gonopods*) beneath their genital opening, which may be equipped with sclerotized hooks or soft sucker-like structures. These are used to take up the sperm packages from the spermatophore (Weygoldt 2000). The exact mechanism of this complex interaction is not entirely clear and remains to be investigated.

2.3.2 Mechanical Mating Plugs in Spiders

In many spiders, males seal the female genital opening after copulation, in order to avoid sperm concurrence and secure their paternity (Uhl et al. 2010). One type of such sealing mechanism is the controlled breakage of the embolus tip (Schneider et al. 2001; Kuntner et al. 2009) or the autotomy of the whole pedipalp (Knoflach and Van Harten 2001), which then stays attached to the epigyne. The embolus tip might stay attached, because it is pressed and locked in the narrow duct, due to the elasticity of both embolus and duct wall material. In single cases, the whole body of the spontaneously dying male serves as a plug against subsequent matings (Knoflach and Benjamin 2003). Its stable attachment might be due to the strong interlocking of the pedipalps with the epigyne.

Another, frequent, mechanism of genital sealing is the application of solidifying substances, which will be discussed in Chap. 8.

2.3.3 *Interlocking of Silk Threads*

Silk threads or bundles of them may attach by interlocking with surface asperities. This may especially play a role in the temporary attachment of bridging threads, which are used by spiders to reach distant sites by air drift (Wolff et al. 2014a). To enhance attachment, the spider toughens the thread by pulling, before it climbs onto the thread. However, the attachment of the bridging thread by interlocking is usually not very durable and especially susceptible to lateral forces acting on the thread by wind. Thus, the spider replaces the bridging thread after its single use by a dragline fastened by the solidifying glue of the piriform glands (see Chap. 8).

Interlocking may also play a significant role in prey attachment to cribellate silk threads, which represent bundles of nano-fibrils, and in wrapping silk (aciniform silk) (see Chap. 4).

2.4 Pincers and Clamps

Pincers and clamps are structures that are actively closed by strong muscles thereby enclosing and attaching an object. In clamping, strong normal forces may act on the enclosed object, which highly enhance friction between the clamp cuticle and the surface of the object. For high friction generation, there are often denticles or ridges on the cuticle of the clamp interior. Clamps work by the flexion of a moveable part against an immobile part: often this function is enhanced by the specific leverage. The closure is performed with the action of strong muscles, the opening may be muscle driven, driven by the internal hemolymph pressure or by the release of stored elastic energy (Sensenig and Shultz 2004).

Pincers and clamps are well-known from crustaceans, but are also an important feature of the Chelicerata (the name of this super-group refers to their scissor like mouth parts), which include the arachnids. The first pair of appendages, the chelicerae, is pincer-like in its ancestral state, which can be found in many arachnids, like scorpions, harvestmen (Video 8.2) and some mites. Pincers have also evolved in some pedipalps, as the case well-known from scorpions, but also in pseudoscorpions and ricinuleids. Clamp-like devices are the raptorial pedipalps of whip-spiders, whip-scorpions and harvestmen, which show strong modifications and enhanced freedom of movement of the femur-patella joints (Wolff et al. 2016b).

2.4.1 *Chelicerae*

In their ancestral state, the chelicerae are composed of three segments and have a pincer-like appearance (Shultz 1990; Sharma et al. 2012). This configuration is found in harvestmen (Fig. 2.4a), scorpions, micro whip-scorpions (Palpigradi) and some mites.

The chelicerae of harvestmen (Fig. 2.4a) are strong claspers and important means of holding prey items. This is especially important, since harvestmen have no venom glands for prey anaesthetization and may have to deal with a long time struggling prey. In various species, the chelicerae are greatly enlarged and massive, and can

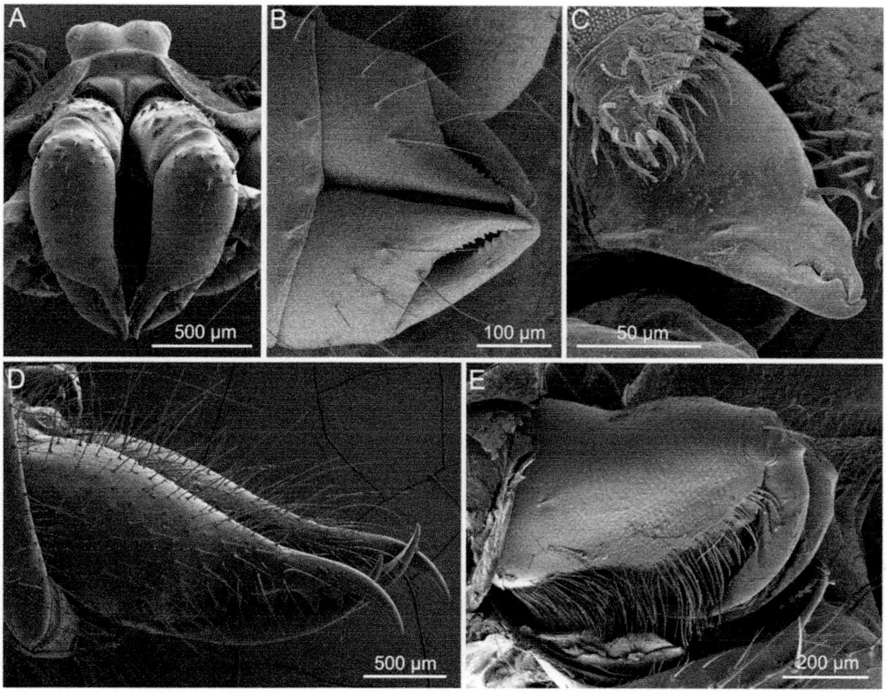

Fig. 2.4 Chelicerae. Chelicerae are the first pair of appendages in arachnids and represent strong clamps for the attachment and dismembering of struggling prey. In its basal state, they are pincer-like, as in (**a**) harvestmen (here: *Dicranopalpus ramosus*, frontal view), (**b**) pseudoscorpions (here: *Neobisum* sp., dorsal view), and (**c**) some mites (here: *Opilioacarus texanus*, lateral view). Derived states are the strong scissor-like chelicerae of (**d**) solifuges (here: juvenile Solpugidae, lateral view), and (**e**) the jackknife like type in the Tetrapulmonata (here: the whip spider *Charinus cubensis*, lateral view)

reach multiple times of the body length. In some species of *Ischyropsalis,* these may be used to grab and break the shells of snails (Verhoeff 1900; Martens 1965), but this was not observed in other species with enlarged chelicerae (J. Martens, pers. comm.). In the Megalopsalidinae (Neopilionidae) (Taylor 2011) and *Rhampsinitus* spp. (Phalangiidae) (Schonhofer 2008), the chelicerae are highly elongated in males, and used in contests (Painting et al. 2015).

In contrast, the chelicerae of spiders only consist of two segments, of which the distal one, the fang, is blade-like. This is because it is used as an injection needle for venomization. However, it can also be used as a clamp (i.e. Video 2.5). In many spiders the basal segment carries tooth-like denticles enhancing friction with a grasped prey item (Foelix 2011). Although the chelicerae are predominantly feeding organs, they may also play a role in courtship and mating, handling (clipping) silk threads, cocoon carrying (in some Pholcidae, Scytodidae and Pisauridae), and burrow digging (Foelix 2011). In the basal lineages of spiders, the Mesothelae and the Mygalomorphae, the fangs of the chelicerae are oriented forwards (orthognath). In the derived spiders, the Araneomorphae, the chelicerae are facing each other (labidognath). This latter apomorphic (derived) trait improves the grasping ability with both chelicerae acting together in a clamp-like manner (Foelix 2011). In some of such spiders, the chelicerae are tilted forwards, like in some *Dysdera* species (Dysderidae), which use them to reach the ventral side of the heavily armored woodlice they prey on (Řezáč et al. 2008). In the Archaeidae, the chelicerae are highly elongated, which is assumed to be an adaptation to preying on other spiders (Legendre 1961; Wood et al. 2012). In the Mecysmaucheniidae and Pararchaeidae, the chelicerae can be widely opened and arrested, and the sudden release of the stored elastic energy is then used in a rapid snap (Vellard 1957; Wood et al. 2016), comparable to trap-jaw ants (Gronenberg et al. 1993). Many Tetragnathidae have elongated, distally diverging chelicerae with long fangs and strong spurs, used to grasp the mating partner (Eberhard and Huber 1998). In some Dictynidae, the males exhibit a curved recess on the inner side of the chelicerae presumably to clasp the female (Chamberlin and Gertsch 1958). A high sexual dimorphism is also known from some salticids, like the common synanthropic *Salticus scenicus* or various ant-mimic species, in which the males with their enormously elongated chelicerae perform contests, in which they try to embrace and clasp each other (Jackson 1982).

Whip spiders and whip-scorpions have jack-knife like chelicerae (Fig. 2.4e), like spiders, however, without venom glands. Hence, they are not primarily used for piercing, but rather for dismembering the prey (own observation). Accordingly, they are equipped with sharp denticles. They are orthognath like in basal spiders and therefore less useful for the usage as a clamp. Thus, whip-spiders use the strong raptorial pedipalps to arrest the prey while feeding (Fig. 2.8a).

The largest and most powerful chelicerae occur in solifuges. They are scissor-like and oriented forwards (Fig. 2.4d). They are equipped with sharp denticles and used to grasp and dismember prey, and in males also to overwhelm the female and insert the spermatophore (Kästner 1933; Bird et al. 2015). The proximal part is massive, providing space for large, strong flexor muscles, which can produce biting forces over 900 kPa (van der Meijden et al. 2012). Prey, which has been grasped with the chelicerae, is barely able to escape this strong clamp. Males exhibit a dorsal cuticular protuberance on the chelicerae, the flagellum, which has been hypothesized to serve as an attachment device for sperm transfer (Kästner 1933). However, its exact function is ambiguous (Lamoral 1975; Punzo 1998b; Bird et al. 2015).

2.4.2 *Chelae*

Scorpions are well-known for their large chelae of the pedipalps, which are used to grasp and secure prey (Fig. 2.5a) or in intraspecific interactions. The pedipalpal tarsus articulates against a tibial outgrowth. The basal section of the tibia is often

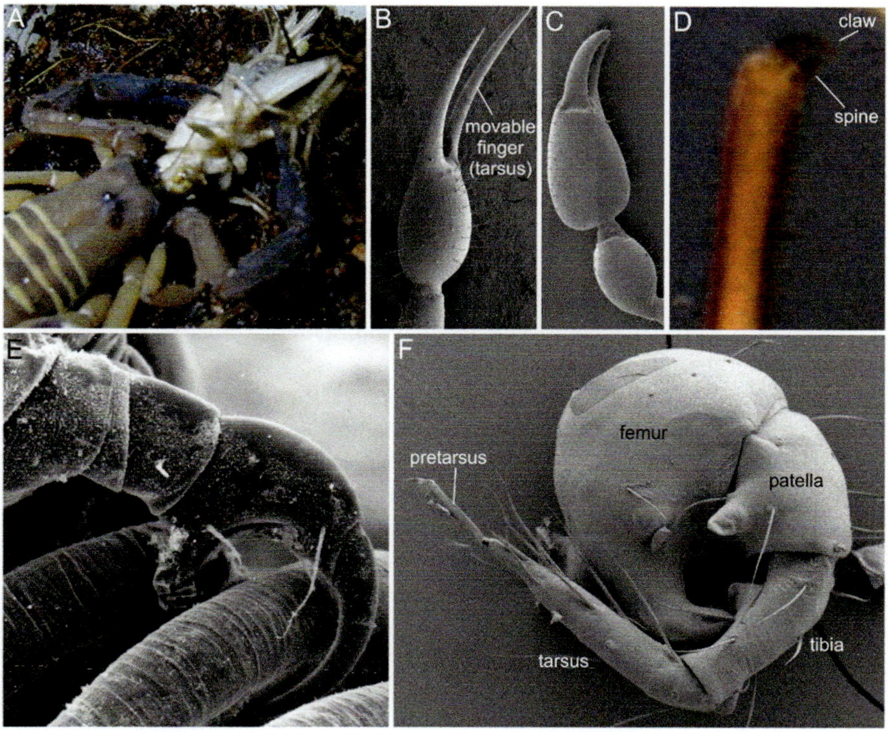

Fig. 2.5 Clamping legs. (**a**) The chelae of the scorpion pedipalp is the most prominent example of efficient clamps. Prey is grabbed and secured while feeding (image by P. Coelho, with kind permission). (**b, c**) Similar chelae are present in the pedipalps of pseudoscorpions. Similarly to scorpions, the proportions of basal part and fingers differ highly between species (here: (**b**) *Neobisium* sp., (**c**) *Atemnus politus*). (**d**) Also the pedipalps of hooded tickspiders (here: *Pseudocellus pachysoma*) carry chelae, although they are comparably small in relation to their body size. (**e**) In parasitic mites, there are often specialized clamping legs, highly adapted to the substrate structure on the host, like in this *Ewingia cenobitae*, which attaches to the gills of freshwater crabs (image adapted from Alberti and Coons 1999, with kind permission by G. Alberti and Wiley). (**f**) Some male mesostigmatid mites have modified front legs to grab and attach to the leg of the female during copula; note the massive femur equipped with a strong spur and the free space, which can perfectly embrace the female's leg

thickened to provide room for thick flexor muscles. Since extensor muscles are lacking, the opening of the chelae is driven by elastic energy stored in a joint sclerite (Sensenig and Shultz 2004). The shape of the chelae can vary from slender and forceps-like to robust and pincer-like, which is correlated with both the prey capture behavior and the use of venom (van der Meijden et al. 2010).

A similar configuration of the pedipalps is present in pseudoscorpions (Fig. 2.5b, c). Like in scorpions, the tibia and tarsus of the pedipalps form a chela, which is used to grasp and manipulate prey items (Beier 1932). In many species, the chela holds a venom gland that opens into the moveable finger (tarsus) and is used to anaesthetize prey (Weygoldt 1969). The chela is also used to attach to phoretic hosts, in order to disperse (Zeh and Zeh 1992). Species that are frequent phorets bear special pointed accessory denticles on the chela (Carl 1994). To withstand the drag forces during flight and grooming, high clamping forces are exerted, which may cause discernible damages to the cuticle of the phoretic host (Carl 1994).

Chelate structures are also present in the pedipalps of Ricinulei, where the pretarsus articulates against a denticulate thorn-like structure on the tarsus (Talarico et al. 2008) (Fig. 2.5d).

2.4.3 Clamping Legs

Predatory arachnids that directly attack and grasp their prey often use their contralateral legs to embrace the prey. Many free hunting spiders (spiders that do not build prey capture webs), like wolf spiders (Lycosidae), wandering spiders (Ctenidae) or lynx spiders (Oxyopidae) often use all legs for prey embracing (capture basket) (Rovner 1980; Eggs et al. 2015) (Video 2.6). Others, like crab spiders (Thomisidae), ground spiders (Gnaphosidae), cribellate hunting spiders (Zoropsidae: *Zoropsis*) or some jumping spiders (Salticidae: *Cyrba*, *Portia*), only use the front legs, which is reflected in front leg enlargement (many Thomisidae, Fig. 2.6a; Video 2.7), dense scopulation (Gnaphosidae, some Salticidae) (see Chap. 3) or spination (Fig. 2.6a–c) (see Sect. 2.2) (Foelix et al. 1984; Grimm 1985; Eggs et al. 2015). Some of these spiders are specialized to overwhelm large or dangerous prey, like other spiders. In such cases, a full capture basket could be risky, since the prey is held close to the body. In the araneophagic Palpimanidae, the front legs are modified into prey capture tools (Fig. 2.7b; Video 2.8). The femur is massively enlarged, containing strong muscles, the femur-patella joint can be highly flexed and the pro-lateral sides of tibia, metatarsus and tarsus are covered by a dense scopula, which produces strong friction on the prey cuticle (see Chap. 3) (Pekár et al. 2011).

Fig. 2.6 (continued) B. Eggs, with kind permission), while others, like *Cupiennius salei* ((**d**) Ctenidae, photo by B. Eggs, with kind permission) fully embrace the prey (*'capture basket'*). Erectable spines (see (**b**) and (**c**) *arrowhead*) enhance attachment by building a barrier and interlocking with prey structures. (**e**, **f**) In parasitic mites living on hairs, like this *Chirodiscoides caviae*, the whole body is shaped to enclose the hair structure, while ridges on the inner cuticle enhance friction (*arrowhead*) (After Alberti and Coons 1999, with kind permission by G. Alberti and Wiley)

Fig. 2.6 The whole body as aclamp. (**a–d**) Many hunting spiders secure prey by embracing it with contra-lateral legs. Some hunting spiders exclusively use the (often enlarged) front legs, like crab spiders ((**a**) Thomisidae, here: *Xysticus* sp., see Video 2.5 for motion, photo/video by N. Ott, with kind permission) or cribellate wandering spiders ((**c**) Zoropsidae, here: *Zoropsis spinimana*, photo by

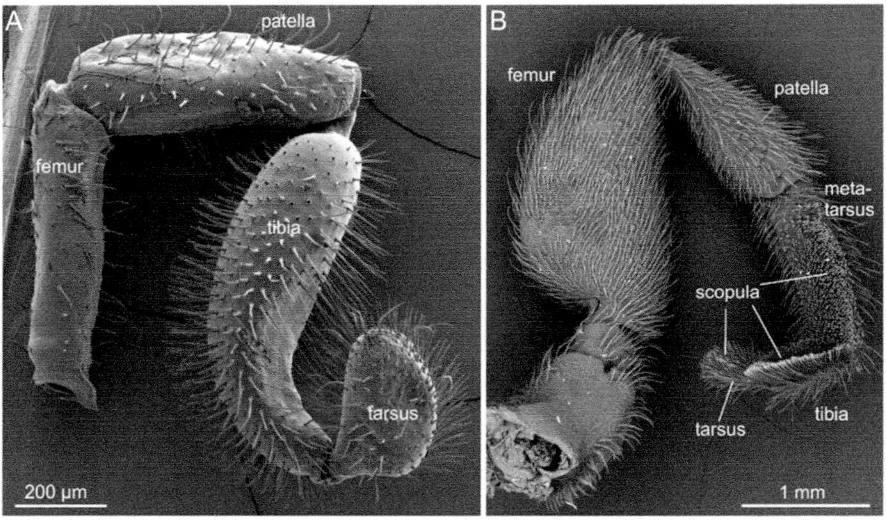

Fig. 2.7 Clamping legs with sticky hairs. (**a**) The tarsus of pedipalps of the harvestman *Sabacon simoni* can be highly flexed against the tibia, which exhibits a concave depression; additionally, strong adhesion and friction are produced by the setae, which extrude a sticky substance. (**b**) The Palpimanidae (here: *Palpimanus gibbulus*) are a family of spiders, which preys on other spiders; their modified front legs are used to grasp and secure the prey; strong friction with the prey cuticle is produced by the dry adhesive setae of the scopula (Also see Video 2.8)

The pedipalps of eupnoid harvestmen show joint modifications that help to exert strong forces to arrest prey or mating partners (Wolff et al. 2016a). In some species, like *Dicranopalpus* spp., these are equipped with long apophyses that may additionally work as a clamp. The pedipalps of most dyspnoid harvestmen are covered by hairs that extrude a sticky substance, and hence work like sticky tentacles to trap prey (Wolff et al. 2014b) (see Chap. 8). To enhance the action of the sticky setae the tarsus can be highly flexed against the tibia, thereby enclosing prey appendages (Wolff et al. 2014b, 2016a, b) (see Video 8.4). In *Sabacon* spp. (see Fig. 2.7a) the flexion is extremely rapid due to an elastic spring element in the tibio-tarsal joint, which is assumed to be an adaptation to outcompete the quick jumping mechanism of the springtail prey (Wolff et al. 2016b).

Some parasitic or commensal mites attach via specialized clamping legs, especially those that live on hairs, feathers or gills, for example *Ewingia coenobitae* (Sarcoptiformes: Acaridae) (Fig. 2.5e) (Alberti and Coons 1999) or *Myobia* spp. (Trombidiformes: Myobiidae) (Paran 1989). In *Chirodiscoides caviae* (Sarcoptiformes: Atopomelidae), a parasite of guinea pigs, the whole body and legs are ventrally concave and exhibit transverse ridges for friction generation in order to attach and move along hairs (Fig. 2.6e, f) (Alberti and Coons 1999).

2.4.4 Raptorial Legs

Whip spiders and whip-scorpions do not have venom glands and thus have to secure the still struggling prey while feeding (Weygoldt 2000). In order to catch and arrest the prey, whip spiders have evolved strong raptorial pedipalps, which are equipped with long curved spines pro-dorsally and pro-ventrally, forming a capture basket (Fig. 2.8a, b). The pedipalpal patella is highly elongated and can be highly flexed against the femur, clamping the prey in between. In prey capture situation, the flexion movement is one of the fastest striking movements in arachnids: in *Phrynus marginemaculatus* it is performed in only about 8 ms, with accelerations of 3–50 ms^{-2} (Santer and Hebets 2009). Such rates are usually too quick to be exerted by muscular contraction, and should involve a kind of elastic spring mechanism (Gronenberg 1996), however, this has not been studied to date in these animals. In

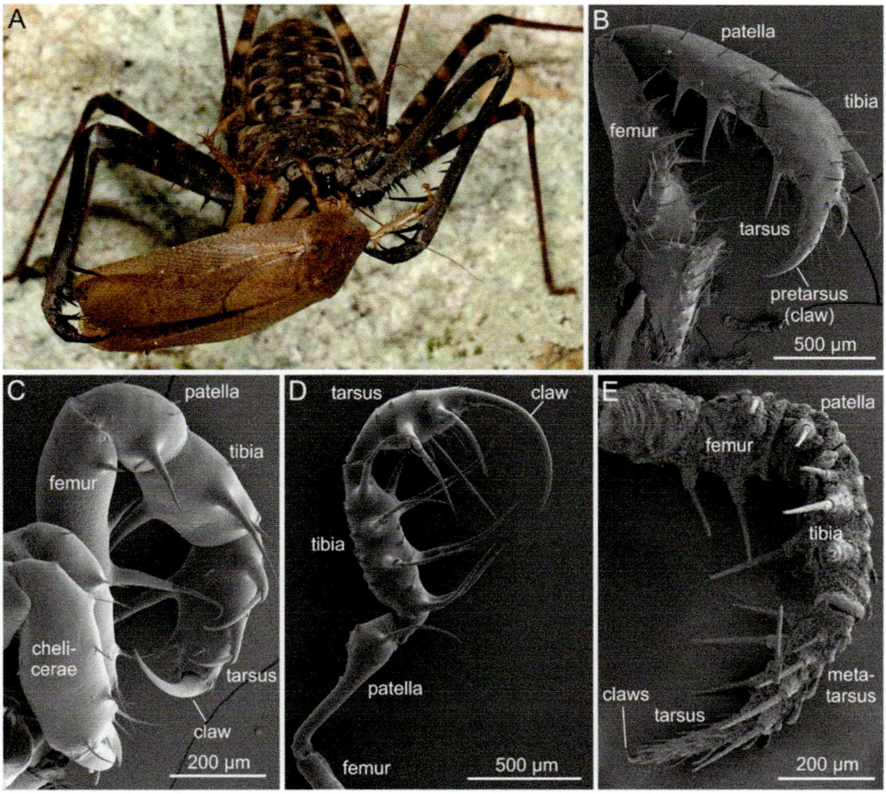

Fig. 2.8 Raptorial legs. (**a, b**) Whip-spiders exhibit raptorial pedipalps for prey capture, which are armed with spines and exhibit a high degree of flexion of the femur-patella joint to exert high clamping forces ((**a**) *Charon grayi*, photo by S. Huber, with kind permission; (**b**) *Charinus cubensis*). (**c, d**) The raptorial pedipalps, typical for harvestmen of the sub-order Laniatores, are not orientated towards each other, but flex parallel to the body axis; various shapes have evolved from stout and massive, like in *Holoscotolemon naturae* ((**c**) Cladonychiidae), or long and slender, like in *Galibrotus* cf. *riedeli* ((**d**) Biantidae). (**e**) In the mite *Caeculus* sp. (Trombidiformes: Caeculidae), the front legs are equipped with strong spines assisting prey capture

male whip-spiders, the raptorial pedipalps also play an important role in ritualized fighting. Males assess their strength by the length of the opened pedipalps, hence an enormous elongation of the pedipalps can be found in males of some species (Weygoldt 2000). In whip-scorpions, the pedipalps are much shorter, more robust and less equipped with spines. In the Schizomida, the pedipalps are oriented forwards, which might be the basal configuration. In the Thelyphonida, the pedipalps are oriented towards each other and can therefore be additionally used to clamp prey in between. The prey is often squeezed by the impact of the strong pedipalps, which assists the feeding process (pers. observation). It is barely possible for the prey to escape this strong grip. Whip-spiders can presumably not exert such high forces with their more slender pedipalps, which work more like a capture basket.

Raptorial legs with striking similarity also evolved in some predatory insects, like mantids, aquatic bugs or mantis flies (Ass 1973; Gorb 1995; Frantsevich and Wang 2009), however, the mechanisms of strike differ greatly in different arthropod groups.

2.4.5 *Cucullus*

Ricinulei bear a unique structure called the *cucullus* ('hood'). It is a dorsal projection, which covers the mouth parts. It can be raised and lowered by muscles and thus also be used in a clamp-like manner. In cooperation with the pedipalps, this is used to capture prey or carry the single eggs (Cooke 1967).

2.5 Expansion Anchors

Some structures may work like expansion anchors. They are pierced and locked within a substrate, like the fangs of spiders or the mouthparts of ticks. The chelicerae of spiders consist of two segments, of which the distal one, the fang, is blade-like and highly pointed, in order to pierce the prey's body. To be able to penetrate the cuticle of arthropod prey, the fang cuticle must be stiffer and tougher, which is achieved by a specific material structure and the incorporation of zinc, calcium and chloride ions (Politi et al. 2012; Bar-On et al. 2014). When pierced into the prey's body and flexed against the basal segment, the fang is locked, and the prey is unable to escape without severe injury. It is, thus, the most important means of prey attachment after successful overwhelming.

In ticks, the main anchoring structures are the barbedhypostome and cheliceral sheath (Ixodida) (Kemp et al. 1982; van der Hammen 1983; Krantz and Walter 2009; Latif et al. 2012) (Fig. 2.9). Ticks also secrete a solidifying substance that additionally affixes the anchorage (see Sect 8.2) (Gregson 1960). In water mite larvae (Hydrachnellae: *Arrenurus* spp.), long claws on the pedipalps cooperate with the penetrating sabre-like chelicerae, to attach to arthrodial membranes of the damselfly larva host (Åbro 1979).

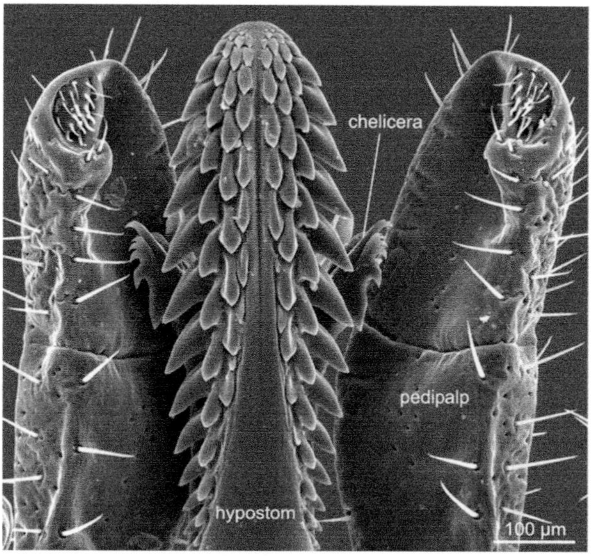

Fig. 2.9 Mouthparts of a tick. The hypostome and chelicerae of the common tick *Ixodes ricinus* can be pierced into the host skin and locked by their barbed backwards-oriented denticles. The barbed chelicerae can be spread, which enhances the anchorage. Additionally, most ticks secrete a solidifying substance, which further enhances the anchorage

2.6 Lock-and-Key

Some structures interlock with highly specific counterpart structures, for example spider genitalia or some spermatophores. This attachment principle is called lock-and-key mechanism. It is assumed to be shaped by sexual selection and plays an important role in speciation processes. However, it is hotly debated, if various interpretations of lock-and-key mechanisms are correct (see Chap. 9).

In male spiders, the pedipalpal tarsi are modified into copulatory organs, the bulbs. These contain a reservoir, into which the sperm is absorbed, and the embolus, a tapering pipe-like protuberance injecting sperm into the female genital tract. In the higher Araneomorphae, the Entelegynae, which include the majority of spider species, the copulatory structures are highly complex and work in a lock-and-key principle together with the female sclerotized genital plate, the epigyne. This interaction is complex, highly specific, and involves, besides mechanical interlocking, a hydraulic inflation mechanism (Huber 1993; Eberhard and Huber 2010). In the Ricinulei, males exhibit secondary copulatory structures in the third leg pair, which may work with a lock-and-key mechanism similar to the pedipalps of entelegyne spiders (Legg 1977).

2.7 Wrapping

Wrapping is a technique, in which an object (more or less cylindrical) is totally enclosed by a thin flexible structure, like a thread. By this mechanism comparably high friction forces can be exerted and maintained at minimal normal load.

2.7.1 Silk Wrapping

Wrapping plays a role in the prey capture of spiders. For this purpose, aciniform silk is utilized (Blackledge and Hayashi 2006). A tight casing is produced in order to restrain the prey. This specific type of attachment presumably works without the intake of any glue. High friction forces may act between single aciniform fibrils due to van der Waals forces at the sites of the very close (real) contact. Prey wrapping behavior has been evolved at least four times independently and may include other types of silk threads and also viscid glue in some taxa (Barrantes and Eberhard 2007). Another mechanism than producing a tight silk casing is the tethering of the prey with dragline silk that is glued onto the substrate by means of piriform gland silk (attachment disks) next to the prey, which is performed by the hunting spider-*Phoneutria boliviensis* (Hazzi 2014). Another type of silk wrapping occurs in nursery spiders (Pisauridae) and some crab spiders (Thomisidae). Here the males temporarily immobilize the female by wrapping its legs with some silk fibers, in order to reduce the risk of being cannibalized prior or during copulation (Anderson and Hebets 2016). In such copulatory wrapping only few threads are used, so that the female can free itself after some time. To some extend tight wrapping is also part of egg casing and cocoon building.

2.7.2 The Prehensile Feet of Harvestmen

Many harvestmen of the Palpatores clade, especially of the Phalangiidae and Sclerosomatidae, have highly elongated legs with an excessive sub-segmentation of the tarsi. These are used as a prehensile organs to grasp corrugated surfaces, leaf edges or plant stems (Kästner 1931; Guffey et al. 2000). The single tarsomeres have a trapeze-like shape, such that there is a high freedom to bend ventrally, whereas the dorsal movement is blocked (Kästner 1931) (Fig. 2.10e). There are no muscle attachment to the tarsomeres, and their joints are indirectly controlled by the pretarsal muscle alone (Kästner 1931). Most harvestmen lack the tarsal depressor muscle, the extension of the tarsi may be driven by an increase of the internal hemolymph pressure, which is adjusted by prosomal muscles (Shultz 2000). Kästner (1931) concluded from his observations that the prehensile tarsi are very effective to navigate through a complex terrain, because they always catch a structure without the necessity to search for foothold. A long-legged harvestman can thus step through the grass and herb vegetation very quickly and without having any adhesive pads. The ventral sides of the prehensile tarsi are often densely covered with setae, which may provide friction and interlocking with minute asperities of natural surfaces (see Sect. 2.2). Microstructures with comparable shape that enhance friction have been found in the modified gripping feet of chameleons (Spinner et al. 2014). In spite of some difference in their dimensions, they might be a functional analogues. In both the harvestman prehensile tarsi and the chameleon foot, the key feature in generating high attachment forces is the embracing of the substrate structure, which

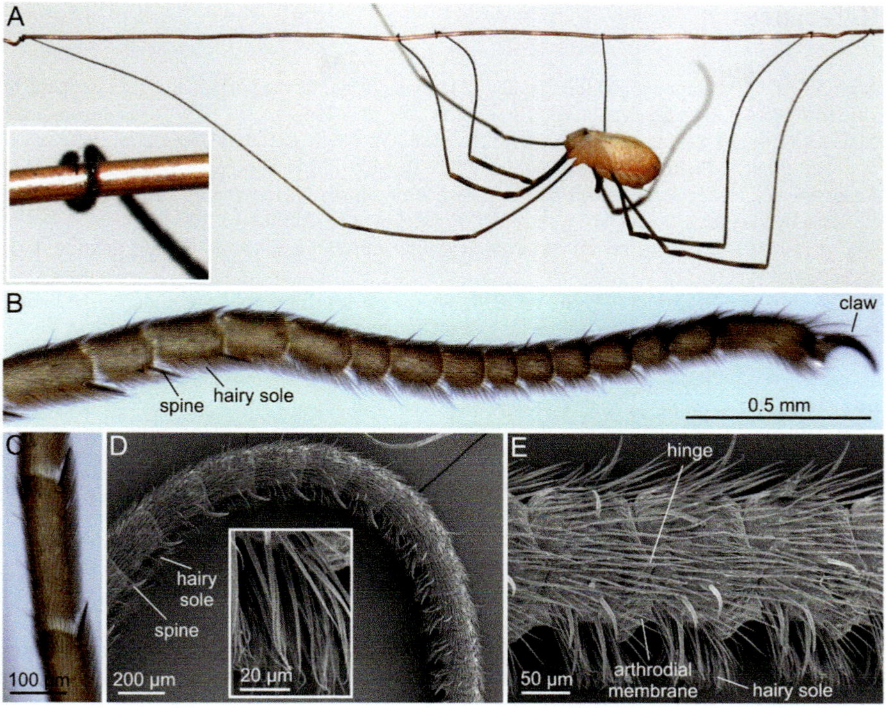

Fig. 2.10 Prehensile feet of harvestmen (Opiliones: Eupnoi). (**a**) The long legged eupnoid harvestmen can easily climb on cylindrical objects like a thin wire, as demonstrated here for *Leiobunum blackwalli* (Sclerosomatidae). Inset shows detail of the left leg I grip. (**b–e**) Distal most parts of hind leg tarsi in *Metagagrella* cf. *minax* (Sclerosomatidae) (**b**, **c**), and *Zacheus crista* (Phalangiidae) (**c**, **d**). *Inset* in (**d**) shows detail of the hairy sole. Note the trapeze-like shape of tarsomeres, permitting a ventral bending, whereas blocking dorsal bending. This causes preferable ventral bending of the tarsal chain (see (**d**)), when the pretarsal muscle is contracting. Note also the gradual decrease of segment length from proximal to distal, and the absence of spines in the distal most segments. The spines are present distally on each segment, near the segment, where they can act best as interlocking devices during contact formation and as levers during contact breakage

permits the redirection of shear forces into normal forces. The very fine and dense setae on the soles may generate high friction even on surfaces with very small roughness scales, at which the efficacy of spatulate setae is highly reduced (see Chap. 3), and may additionally work as a fibrillary adhesive (see Chap. 4). In some harvestmen species, short but rigid spines are additionally included in the dense hairy sole (Fig. 2.10b–d): these spines may enhance the efficacy on more planar surfaces, like tree stems. Some Laniatores also exhibit highly sub-segmented tarsi, however, never to that extend like in the cases of Phalangiidae or Sclerosomatidae. In contrast, these harvestmen still possess both pretarsal muscles (Proud and Felgenhauer 2013) and the tarsi are more flexible in the dorsal than in the ventral direction. Hence, they can hardly wrap around plant structures. In the harvestmen legs, a system of pulleys guides the tendons to keep proper motion control over the high amount of adesmatic joints (Proud and Felgenhauer 2013).

References

Åbro A (1979) Attachment and feeding devices of water-mite larvae (*Arrenurus* spp.) parasitic on damselflies (Odonata, Zygoptera). Zool Scr 8(1–4):221–234

Alberti G, Coons L (1999) Acari: mites. In: Harrison FW, Foelix RF (eds) Microscopic anatomy of invertebrates: Chelicerate Arthropoda, vol 8. Wiley-Liss, New York, pp 515–1265

Anderson AG, Hebets EA (2016) Benefits of size dimorphism and copulatory silk wrapping in the sexually cannibalistic nursery web spider, *Pisaurina mira*. Biol Lett 12(2):20150957

Ass M (1973) Die Fangbeine der Arthropoden, ihre Entstehung, Evolution und Funktion. Deut Entomol Z 20(1-3):127–152

Bar-On B, Barth FG, Fratzl P, Politi Y (2014) Multiscale structural gradients enhance the biomechanical functionality of the spider fang. Nat Commun 5:3894

Barrantes G, Eberhard WG (2007) The evolution of prey-wrapping behaviour in spiders. J Nat Hist 41(25–28):1631–1658

Beier M (1932) 5. Ordnung der Arachnida: Pseudoscorpionidea = Afterscorpione. In: Kükenthal W, Krumbach T (eds) Handbuch der Zoologie: Bd. 3, 2. Hälfte. Chelicerata, vol 3. W. de Gruyter & Co., pp 117–192

Bertani R, Guadanucci JPL (2013) Morphology, evolution and usage of urticating setae by tarantulas (Araneae: Theraphosidae). Zoologia (Curitiba) 30(4):403–418

Bird TL, Wharton RA, Prendini L (2015) Cheliceral morphology in Solifugae (Arachnida): primary homology, terminology, and character survey. Bull Am Mus Nat Hist 916(1):1–356

Blackledge TA, Hayashi CY (2006) Silken toolkits: biomechanics of silk fibers spun by the orb web spider *Argiope argentata* (Fabricius 1775). J Exp Biol 209(13):2452–2461

Bolton SJ, Bauchan GR, Ochoa R, Pooley C, Klompen H (2015) The role of the integument with respect to different modes of locomotion in the Nematalycidae (Endeostigmata). Exp Appl Acarol 65(2):149–161

Carl M (1994) Injurious effects on the exoskeleton of *Musca domestica* L. (Diptera) of phoresy by *Lamprochernes nodosus* (Schrank) (Pseudoscorpiones, Chernetidae) and the possible functional significance of accessory teeth on the chelal fingers. Bull Br Arachnol Soc 9(7):246–248

Chamberlin RV, Gertsch WJ (1958) The spider family Dictynidae in America north of Mexico. Bull Am Mus Nat Hist 111(1):1–152

Cooke JA (1967) Obsevations on the biology of Ricinulei (Arachnida) with descriptions of two new species of *Cryptocellus*. J Zool 151(1):31–42

Dai Z, Gorb SN, Schwarz U (2002) Roughness-dependent friction force of the tarsal claw system in the beetle *Pachnoda marginata* (Coleoptera, Scarabaeidae). J Exp Biol 205(16):2479–2488

Ditsche-Kuru P, Barthlott W, Koop JH (2012) At which surface roughness do claws cling? Investigations with larvae of the running water mayfly *Epeorus assimilis* (Heptageniidae, Ephemeroptera). Zoology 115(6):379–388

Dunlop JA (2002) Character states and evolution of the chelicerate claws. In: Toft S, Scharff N (eds) European arachnology 2000. Aarhus University Press, Aarhus, pp 345–354

Eberhard WG, Huber BA (1998) Courtship, copulation, and sperm transfer in *Leucauge mariana* (Araneae, Tetragnathidae) with implications for higher classification. J Arachnol 26(3):342–368

Eberhard WG, Huber BA (2010) Spider genitalia. The evolution of primary sexual characters in animals. Oxford University Press, New York, pp 249–284

Eggs B, Wolff JO, Kuhn-Nentwig L, Gorb SN, Nentwig W (2015) Hunting without a web: how lycosoid spiders subdue their prey. Ethology 121(12):1166–1177

Foelix RF (1970) Structure and function of tarsal sensilla in the spider *Araneus diadematus*. J Exp Zool 175(1):99–123

Foelix RF (2011) Biology of spiders, 3rd edn. Oxford University Press, Oxford / New York

Foelix R, Jackson RR, Henksmeyer A, Hallas S (1984) Tarsal hairs specialized for prey capture in the salticid *Portia*. Rev Arachnol 5:329–334

Foelix R, Rechenberg I, Erb B, Joel A-C (2016) Zum Sandtransport der „Radlerspinne“*Cebrennus rechenbergi* Jäger, 2014. Arachne 20(6):14–21

Fontaine A, Olsen N, Ring R, Singla C (1991) Cuticular metal hardening of mouthparts and claws of some forest insects of British Columbia. J Entomol Soc Brit Columbia 88:45–55
Forster RR, Platnick NI, Gray MR (1987) A review of the spider superfamilies Hypochiloidea and Austrochiloidea (Araneae, Araneomorphae). Bull Am Mus Nat Hist 185:1–116
Frantsevich L, Wang W (2009) Gimbals in the insect leg. Arthropod Struct Dev 38(1):16–30
Gawryszewski FM (2014) Evidence suggests that modified setae of the crab spiders *Stephanopis* spp. fasten debris from the background. Zoomorphology 133(2):205–215
Gillespie RG (1991) Predation through impalement of prey: the foraging behavior of *Doryonychus raptor* (Araneae: Tetragnathidae). Psyche 98(4):337–350
Gorb SN (1995) Design of the predatory legs of water bugs (Hemiptera: Nepidae, Naucoridae, Notonectidae, Gerridae). J Morphol 223(3):289–302
Gorb SN (1999) Evolution of the dragonfly head-arresting system. Proc R Soc Lond B 266(1418):525–535
Gray M (1983) Male of *Progradungula carraiensis* Forster and Gray (Araneae, Gradungulidae) with observations on the web and prey capture. Proc Linnean Soc NSW 107:51–58
Gregson JD (1960) Morphology and functioning of the mouthparts of *Dermacentor andersoni* Stiles. Acta Trop 17(1):48–49
Grimm U (1985) Die Gnaphosidae Mitteleuropas (Arachnida, Araneae). Abh naturwiss Ver Hamburg (NF) 26:1–318
Griswold CE, Audisio T, Ledford JM (2012) An extraordinary new family of spiders from caves in the Pacific Northwest (Araneae, Trogloraptoridae, new family). Zookeys 215:77
Gronenberg W (1996) Fast actions in small animals: springs and click mechanisms. J Comp Physiol A 178(6):727–734
Gronenberg W, Tautz J, Hölldobler B (1993) Fast trap jaws and giant neurons in the ant *Odontomachus*. Science 262(5133):561–563
Guffey C, Townsend VR, Felgenhauer BE (2000) External morphology and ultrastructure of the prehensile region of the legs of *Leiobunum nigripes* (Arachnida, Opiliones). J Arachnol 28(2):231–236
Hazzi NA (2014) Natural history of *Phoneutria boliviensis* (Araneae: Ctenidae): habitats, reproductive behavior, postembryonic development and prey-wrapping. J Arachnol 42(3):303–310
Heethoff M, Koerner L (2007) Small but powerful: the oribatid mite *Archegozetes longisetosus* Aoki (Acari, Oribatida) produces disproportionately high forces. J Exp Biol 210(17):3036–3042
Hill DE (1977) The pretarsus of salticid spiders. Zool J Linnean Soc 60(3):319–338
Hill D (1978) Function of the pretarsus in living *Phidippus regius*. Peckhamia 1:70–71
Huber BA (1993) Genital mechanics and sexual selection in the spider *Nesticus cellulanus* (Araneae: Nesticidae). Can J Zool 71(12):2437–2447
Jackson R (1982) The biology of ant-like jumping spiders: intraspecific interactions of *Myrmarachne lupata* (Araneae, Salticidae). Zool J Linnean Soc Lond 76(4):293–319
Jacob A, Gantenbein I, Braunwalder ME, Nentwig W, Kropf C (2004) Morphology and function of male genitalia (spermatophores) in *Euscorpius italicus* (Euscorpiidae, Scorpiones): complex spermatophore structures enable safe sperm transfer. J Morphol 260(1):72–84
Kästner A (1931) Biologische Beobachtungen an Phalangiiden. Zool Anz 95:293–302
Kästner A (1933) 6. Ordnung der Arachnida. Solifugae Sundvall Walzenspinnen. In: Kükenthal W, Krumbach T (eds) Handbuch der Zoologie, vol 3, vol 2. de Gruyter & Co., Berlin, pp 193–299
Kemp D, Stone B, Binnington K (1982) Tick attachment and feeding: role of the mouthparts, feeding apparatus, salivary gland secretions, and the host response. In: Obenchain FD, Galun R (eds) Physiology of ticks, vol 1. Pergamon Press, Oxford, pp 119–167
Kew H (1930) On the spermatophores of the pseudoscorpions *Chthonius* and *Neobisium*. Proc Zool Soc London 1930:253–256
Klann AE (2009) Histology and ultrastructure of solifuges. Comparative studies of organ systems of solifuges (Arachnida, Solifugae) with special focus on functional analyses and phylogenetic interpretations. Ph.D thesis, Ernst-Moritz-Arndt-University Greifswald, Greifswald, 236pp.

Published online https://www.deutsche-digitale-bibliothek.de/binary/ODU3XEEXCIZGVHOOHIRD7XTONYTTEVS7/full/1.pdf
Knoflach B, Benjamin SP (2003) Mating without sexual cannibalism in *Tidarren sisyphoides* (Araneae, Theridiidae). J Arachnol 31(3):445–448
Knoflach B, Van Harten A (2001) Tidarren argo sp. nov. (Araneae: Theridiidae) and its exceptional copulatory behaviour: emasculation, male palpal organ as a mating plug and sexual cannibalism. J Zool 254(04):449–459
Krantz GW, Walter DE (2009) A manual of acarology, 3rd edn. Texas Tech University Press, Lubbock, 816pp
Kuntner M, Kralj-Fišer S, Schneider J, Li D (2009) Mate plugging via genital mutilation in nephilid spiders: an evolutionary hypothesis. J Zool 277(4):257–266
Labonte D, Federle W (2015) Scaling and biomechanics of surface attachment in climbing animals. Philos Trans R Soc London B 370(1661):20140027
Lamoral B (1975) The structure and possible function of the flagellum in four species of male solifuges of the family Solpugidae. In: Proceedings of the 6th international arachnological congress, 1975. pp 136–141
Latif AA, Putterill JF, De Klerk DG, Pienaar R, Mans BJ (2012) *Nuttalliella namaqua* (Ixodoidea: Nuttalliellidae): first description of the male, immature stages and re-description of the female. Plos One 7(7):e41651
Legendre R (1961) Études sur les Archaea (Aranéides). II. La capture des proies et la prise de nourriture. Bull Soc Zool France 86:316–319
Legg G (1973) Spermatophore formation in the pseudoscorpion *Chthonius ischnocheles* (Chthoniidae). J Zool 170(3):367–394
Legg G (1977) Sperm transfer and mating in *Ricinoides hanseni* (Ricinulei: Arachnida). J Zool 182(1):51–61
Martens J (1965) Verbreitung und Biologie des Schneckenkankers *Ischyropsalis hellwigi*. Nat Mus 95(4):143–149
Meyer MKPS, Ueckermann EA (1987) A taxonomic study of some Anystidae (Acari: Prostigmata). Entomology Mem Dep Agric Wat Supply Repub S Afr 68:1–37
Painting CJ, Probert AF, Townsend DJ, Holwell GI (2015) Multiple exaggerated weapon morphs: a novel form of male polymorphism in harvestmen. Sci Rep 5:16368
Paran TP (1989) Evolution of leg I as an organ of anchorage in *Myobia (Myobia) muris-musculi* (Acari: Prostigmata) and other selected myobiids. In: Channabasavanna GP, Viraktamath CA (eds) Progress in acarology. Brill, Leiden, pp 229–240
Pekár S, Šobotník J, Lubin Y (2011) Armoured spiderman: morphological and behavioural adaptations of a specialised araneophagous predator (Araneae: Palpimanidae). Naturwissenschaften 98(7):593–603
Peretti AV (2004) Functional morphology of spermatophores and female genitalia in bothriurid scorpions: genital courtship, coercion and other possible mechanisms. J Zool 261(2):135–153
Pinto-da-Rocha R, Machado G, Giribet G (2007) Harvestmen : the biology of Opiliones. Harvard University Press, Cambridge, MA
Platnick NI (1976) Notes on the spider genus *Doliomalus* (Araneae, Gnaphosoidea). Revue Zoologique Africaine 90:975–983
Politi Y, Priewasser M, Pippel E, Zaslansky P, Hartmann J, Siegel S, Li C, Barth FG, Fratzl P (2012) A spider's fang: how to design an injection needle using chitin-based composite material. Adv Funct Mater 22(12):2519–2528
Proud DN, Felgenhauer BE (2013) The harvestman tarsus and tarsal flexor system with notes on appendicular sensory structures in laniatores. J Morphol 274(11):1216–1229
Pugh P, King P, Fordy M (1987) Ambulacral structure in the terrestrial moiety of the intertidal Acari, and its relationship with the lifestyle of the Acari. Acarologia 28:3–13
Punzo F (1998a) The effects of reproductive status on sprint speed in the solifuge, *Eremobates marathoni* (Solifugae, Eremobatidae). J Arachnol 26(1):113–116

Punzo F (1998b) Functional anatomy and physiology. In: The biology of camel-spiders. Springer, Boston, pp 11–44
Ramírez MJ (2014) The morphology and phylogeny of dionychan spiders (Araneae, Araneomorphae). Bull Am Mus Nat Hist 390:1–374
Ramírez MJ, Grismado CJ, Labarque FM, Izquierdo MA, Ledford JM, Miller JA, Haddad CR, Griswold CE (2014) The morphology and relationships of the walking mud spiders of the genus *Cryptothele* (Araneae: Zodariidae). Zool Anz 253(5):382–393
Raven RJ (1985) The spider infraorder Mygalomorphae (Araneae): cladistics and systematics. Bull Am Mus Nat Hist 182(1):1–182
Řezáč M, Pekár S, Lubin Y (2008) How oniscophagous spiders overcome woodlouse armour. J Zool 275(1):64–71
Roewer CF (1923) Die Weberknechte der Erde. Gustav Fischer, Jena
Roewer CF (1935) Opiliones (Fünfte série). Zugleich eine Revision aller bisher bekannten europäischen Laniatores. Bioespeleologica LXII. Arch Zool Exp Gén 78 (1):1–96
Rovner JS (1980) Morphological and ethological adaptations for prey capture in wolf spiders (Araneae, Lycosidae). J Arachnol 8(3):201–215
Rovner JS, Higashi GA, Foelix RF (1973) Maternal behavior in wolf spiders: the role of abdominal hairs. Science 182(4117):1153–1155
Santer RD, Hebets EA (2009) Prey capture by the whip spider *Phrynus marginemaculatus* CL Koch. J Arachnol 37(1):109–112
Schneider JM, Thomas ML, Elgar MA (2001) Ectomised conductors in the golden orb-web spider, *Nephila plumipes* (Araneoidea): a male adaptation to sexual conflict? Behav Ecol Sociobiol 49(5):410–415
Schofield R, Nesson M, Richardson KA, Wyeth P (2003) Zinc is incorporated into cuticular "tools" after ecdysis: the time course of the zinc distribution in "tools" and whole bodies of an ant and a scorpion. J Insect Physiol 49(1):31–44
Schonhofer AL (2008) On harvestmen from the Soutpansberg, South Africa, with description of a new species of Monomontia (Arachnida: Opiliones). Afr Invertebr 49(2):109–126
Sensenig AT, Shultz JW (2004) Elastic energy storage in the pedipalpal joints of scorpions and sun-spiders (Arachnida, Scorpiones, Solifugae). J Arachnol 32(1):1–10
Sharma PP, Schwager EE, Extavour CG, Giribet G (2012) Evolution of the chelicera: a dachshund domain is retained in the deutocerebral appendage of Opiliones (Arthropoda, Chelicerata). Evol Dev 14(6):522–533
Shultz JW (1990) Evolutionary morphology and phylogeny of Arachnida. Cladistics 6(1):1–38
Shultz JW (2000) Skeletomuscular anatomy of the harvestman *Leiobunum aldrichi* (Weed, 1893) (Arachnida : Opiliones : Palpatores) and its evolutionary significance. Zool J Linnean Soc 128(4):401–438
Spagna JC, Goldman DI, Lin P-C, Koditschek DE, Full RJ (2007) Distributed mechanical feedback in arthropods and robots simplifies control of rapid running on challenging terrain. Bioinspir Biomim 2(1):9
Spinner M, Westhoff G, Gorb SN (2014) Subdigital setae of chameleon feet: friction-enhancing microstructures for a wide range of substrate roughness. Sci Rep 4:5481
Talarico G, Palacios-Vargas J, Alberti G (2008) The pedipalp of *Pseudocellus pearsei* (Ricinulei, Arachnida) – ultrastructure of a multifunctional organ. Arthropod Struct Dev 37(6):511–521
Taylor CK (2011) Revision of the genus *Megalopsalis* (Arachnida: Opiliones: Phalangioidea) in Australia and New Zealand and implications for phalangioid classification. Zootaxa 2773:1–65
Uhl G, Nessler SH, Schneider JM (2010) Securing paternity in spiders? A review on occurrence and effects of mating plugs and male genital mutilation. Genetica 138(1):75–104
van der Hammen L (1983) Notes on the comparative morphology of ticks (Anactinotrichida: Ixodida). Zoologische Mededelingen 57:209–242
van der Meijden A, Herrel A, Summers A (2010) Comparison of chela size and pincer force in scorpions; getting a first grip. J Zool 280(4):319–325

van der Meijden A, Langer F, Boistel R, Vagovic P, Heethoff M (2012) Functional morphology and bite performance of raptorial chelicerae of camel spiders (Solifugae). J Exp Biol 215(19):3411–3418

Vellard J (1957) Études fuégiennes, Premiere note: araigneés de la Terre de Feu. Institut Française d'Études Andines 6:119–146

Verhoeff C (1900) Zur Biologie von *Ischyropsalis*. Zool Anz 23:106–107

Vitzthum H (1943) Dr. Bronns Klassen und Ordnungen des Tierreichs Bd. 5, Abt. 4, Buch 5: Acari. Akademische Verlagsgesellschaft, Leipzig

Weygoldt P (1966) Mating behavior and spermatophore morphology in the pseudoscorpion *Dinocheirus tumidus* banks (Cheliferinea, Chernetidae). Biol Bull 130(3):462–467

Weygoldt P (1969) Biology of pseudoscorpions. Harvard University Press, Cambridge, 159pp

Weygoldt P (1988) Sperm transfer and spermatophore morphology in the whip scorpion *Thelyphonus linganus* (Arachnida: Uropygi: Thelyphonidae). J Zool 215(1):189–196

Weygoldt P (2000) Whip spiders (Chelicerata: Amblypygi): their biology, morphology and systematics. Apollo Books, Stenstrup, 160pp

Weygoldt P, Huber S (2013) Sperm transfer and maternal care in *Thelyphonus* cf *caudatus* from Lombok, Indonesia (Arachnida, Uropygi, Thelyphonida). Zool Anz 252(3):348–349

Wolff JO, Nentwig W, Gorb SN (2013) The great silk alternative: multiple co-evolution of web loss and sticky hairs in spiders. Plos One 8(5):e62682

Wolff JO, Schneider JM, Gorb SN (2014a) How to pass the gap – functional morphology and biomechanics of spider bridging threads. In: Asakura T, Miller T (eds) Biotechnology of silk, vol 5. Springer, Dordrecht, pp 165–177

Wolff JO, Schönhofer AL, Schaber CF, Gorb SN (2014b) Gluing the 'unwettable': soil-dwelling harvestmen use viscoelastic fluids for capturing springtails. J Exp Biol 217(19):3535–3544

Wolff JO, Huber SJ, Gorb SN (2015a) How to stay on mummy's back: morphological and functional changes of the pretarsus in arachnid postembryonic stages. Arthropod Struct Dev 44(4):301–312

Wolff JO, Seiter M, Gorb SN (2015b) Functional anatomy of the pretarsus in whip spiders (Arachnida, Amblypygi). Arthropod Struct Dev 44(6):524–540

Wolff JO, Schönhofer AL, Martens J, Wijnhoven H, Taylor C, Gorb SN (2016a) The evolution of pedipalps and glandular hairs as predatory devices in harvestmen (Arachnida, Opiliones). Zool J Linnean Soc 177:558–601

Wolff JO, Martens J, Schönhofer AL, Gorb SN (2016b) Evolution of hyper-flexible joints in sticky prey capture appendages of harvestmen (Arachnida, Opiliones). Organism Div Ev in press/ online pre-pub. doi:10.1007/s13127-016-0278-2

Wood HM, Griswold CE, Gillespie RG (2012) Phylogenetic placement of pelican spiders (Archaeidae, Araneae), with insight into evolution of the "neck" and predatory behaviours of the superfamily Palpimanoidea. Cladistics 28(6):598–626

Wood HM, Parkinson DY, Griswold CE, Gillespie RG, Elias DO (2016) Repeated evolution of power-amplified predatory strikes in trap-jaw spiders. Curr Biol 26(8):1057–1061

Zeh DW, Zeh JA (1992) Failed predation or transportation? Causes and consequences of phoretic behavior in the pseudoscorpion *Dinocheirus arizonensis* (Pseudoscorpionida: Chernetidae). J Insect Behav 5(1):37–49

Chapter 3
Tape- and Spatula-Shaped Microstructures

Abstract Spatula-like microstructures recruit adhesive forces by generating a close contact with the substrate due to elastic deformation. They usually occur at the tip of bendable hair-like structures, such as setae or microtrichia, whose shafts additionally contribute to the overall conformability of the attachment device. Such spatulate hairs have analogously evolved in the feet of different animal groups and are efficient means of dynamic attachment. The adhesive setae of geckoes are the best-known example of a dry adhesive, based on this principle, but very similarly sized and shaped structures are also present in many hunting spiders and various mites of the order Trombidiformes. Fluid supplemented setae with single, broad spatulae, as known from beetles and flies, are analogously present in some harvestmen of the order Laniatores and in hooded tickspiders (Ricinulei). A unique type of an adhesive pad with micro-spatulae as direct outgrowths of the epicuticle instead of the flattened tips of hairs is represented by the arolium of whip-spiders of the families Charinidae and Charontidae. These pads show contact mechanics and a dry adhesive mechanism comparable to that of the hairy adhesive pads of spiders, and are thus not comparable with other arolia having smooth surfaces. A special type of tape-like adhesive structure is the dragline silk of the brown recluse spider (*Loxosceles* spp.), which is tacky due to its thin, flattened shape.

3.1 Introduction

Among arthropods there are numerous examples of parallel evolution of reversible glue-less adhesive surfaces that are based on tape-like contact elements (Gorb 2001; Gorb and Beutel 2001; Arzt et al. 2003). Such adhesives are most abundant in devices serving dynamic attachment during locomotion, represented by so-called hairy adhesive foot pads (fibrillar pads), which are composed of spatulate, so-called, tenent setae. The physical principles of such hairy adhesive pads have previously been reviewed by Gorb (2001) and Federle (2006). The spatula is the terminal contact element, which generates an intimate contact with the substrate, such that

Electronic supplementary material The online version of this chapter (doi:10.1007/978-3-319-45713-0_3) contains supplementary material, which is available to authorized users Videos can also be accessed at http://www.springerimages.com/videos/978-3-319-45712-3.

J.O. Wolff, S.N. Gorb, *Attachment Structures and Adhesive Secretions in Arachnids*, Biologically-Inspired Systems 7,
DOI 10.1007/978-3-319-45713-0_3

inter-molecular attractive interactions may be generated. It can be modeled as a soft, elastic and thin tape (Varenberg et al. 2010; Filippov et al. 2011). In a tape that is in contact with a smooth surface, the work required to peel it off determines the robustness of the attachment. If the tape is very thin and soft, the initial force necessary to overcome the tape adhesion depends primarily on the width of the tape, because the stress is then acting locally in the delamination zone (peeling edge) (Kendall 1975).

Due to the close contact with the substrate a tape- or spatula-like terminal structure can adhere by short ranging intermolecular forces, like van-der-Waals forces (dry adhesion), or by the capillary and/or viscous forces of thin films of fluid secretions (wet adhesion). Because the single setae are elastic and can deform rather independently from the neighboring ones, the pad can deform very locally and adapt well to the surface topography of a rough substrate. Further, the adhesion failure of a single spatula does barely affect the bonding of the neighboring setae. As the initiation of delamination demands more energy than its propagation, it is assumed that it is beneficial to split the contact area into several sub-contacts, such that a crack is stopped at each gap and has to be repeatedly initiated at each sub-contact (Peressadko and Gorb 2004; Chung and Chaudhury 2005). This observation led to the theory that the adhesion of a fibrillar pad does less depend on its contact area, but primarily on the sum of the perimeter (Arzt et al. 2003) or on the sum of the width of all contacting spatulae (peeling line length) (Varenberg et al. 2010). The contact splitting theory is also applicable, if the spatulae are fluid mediated, because then the capillary and viscous forces depend on the circumference of the contacts (Qian and Gao 2006).

An overall tendency in the structure of such hairy/fibrillar adhesive pads is an increased degree of contact splitting with an increasing body mass, with the largest animals, the geckoes, exhibiting the smallest spatulae (width 0.1–0.4 μm), standing in the highest density (about 4,000,000 spatulae per mm^2) (Arzt et al. 2003). With their increasing density, the spatulae tend to stick to each other (self-matting), which may highly reduce the functionality of the pad (Jagota and Bennison 2002; Spolenak et al., 2005; Federle 2006). Hence, such high densities of individual contacts, as in geckoes, can only be achieved by setal branching: Each seta splits up at its tip or bears numerous microtrichia which end in a spatula. Whereas the relationship between an increasing spatular density with an increasing body mass holds between different lineages of animals that analogously evolved hairy adhesive pads, this is not necessarily the case for animals within the same lineage, and is thus controversially discussed (Peattie and Full 2007; Labonte et al. 2016). Furthermore, the finding may be biased by the choice of included taxa, as some tiny mites exhibit spatulae very similar to large spiders and geckoes (see Sect 3.3 and Chap. 10).

The adhesion by spatulate adhesive setae needs an initial, directed shear movement (activation), by which the spatula is spread onto the substrate (Autumn et al. 2006; Filippov et al. 2011). The setae are usually orientated in such a way that this movement is proximal (towards the body) (Niederegger and Gorb 2006; Autumn et al. 2006; Wolff and Gorb 2013). By distal movement, the spatula turns over due to its flexibility, and looses the contact (deactivation). The rear part is often corrugated or equipped with nubs or short protuberances that prevent generation of a high contact area (Gorb 2001; Niederegger et al. 2002; Haas and Gorb, 2004; Federle, 2006). By this anisotropy, the animal can quickly and effortlessly switch between high and low adhesion, which is crucial for fast locomotion.

3.2 Spatulate Setae

Hairy adhesive pads composed of setae with spatula-like tips are wide-spread among insects (Gorb 2001; Gorb and Beutel 2001), but rather rare among arachnids. The spatula width of such setae ranges from 1 to 10 μm (Peattie and Full 2007). This is usually too large to generate a sufficient contact area on natural surfaces, which exhibit always a certain degree of roughness, hence in most cases a thin film of fluid secretion (usually lipids or oil-water emulsions) is released into the contact, filling the interspaces and acting as a bonding agent by recruiting viscous and capillary forces (Gorb 1998; Federle 2006). Still, such large spatulae are more affected by fine surface roughness than the smaller ones built by microtrichia (see Sect. 3.3) (Wolff and Gorb 2012c).

Among arachnids, spatulate setae only occur in the distal tarsomeres of some tropical harvestmen of the sub-order Laniatores and in hooded tickspiders (Ricinulei). In such species of harvestmen, the hairy soles of the tarsal segments of both posterior leg pairs are discernibly modified, and hence called *scopulae* (Roewer 1923). This character has independently evolved in the Ibaloninae (Podoctidae), Acrobuninae (Epedanidae), Heterostygninae (Stygnidae), Biantidae and Samoidae (Roewer 1923; Pinto-da-Rocha et al. 2007). In all cases, the structure of the scopula setae is comparable. The setae have ribbon-like tips, which may be distally broadened (spatulae), and are assumed to work as adhesive devices (Rambla 1990). However, it has never been tested if these harvestmen are capable to walk on smooth surfaces, and it is unknown how these presumed attachment devices are used in real behavioral situations. Some of these species, like the Biantidae, are nearly exceptionally found at the ground, but knowledge on the biology of these harvestmen is very sparse. In *Hinzuanius flaviventris* (Biantidae), we studied both adults and nymphs and found that the scopula is not developed in the nymphs, which bear the arolium (see Chap. 7). In the Biantidae, the setae are spatulate only in the distal most tarsal segment, which is slightly broadened (Fig. 3.1d). There are no obvious morphological differences between the scopulae of the old world Biantinae (studied: *H. flaviventris*, Fig. 3.1c, f, g) and new world Stenostygninae (studied: *Galibrotus* cf. *riedeli*, Fig. 3.1d, h, i). The setal rear is nubby and at the level of the spatula equipped with a brim (Fig. 3.1i), presumably to prevent a substrate contact of these parts, as known from similar adhesive setae in beetles (Gorb 2001). In *Metacrobunus* sp. (Epedanidae), spatulate setae occur on the two distal segments, which are not significantly broadened (Fig. 3.1b). The scopula is less dense than that in the Biantidae, the setae are much longer and bear much larger spatulae (Fig. 3.1a, b, e). The setae are band-like flattened, get gradually broadened towards the tip, but end more or less pointed, like a leaf (Fig. 3.1e). The back is not nubby. These differences underline the independent evolution of the scopula. In *Dibunus similis*, belonging to the same family (Epedanidae), the scopula is absent.

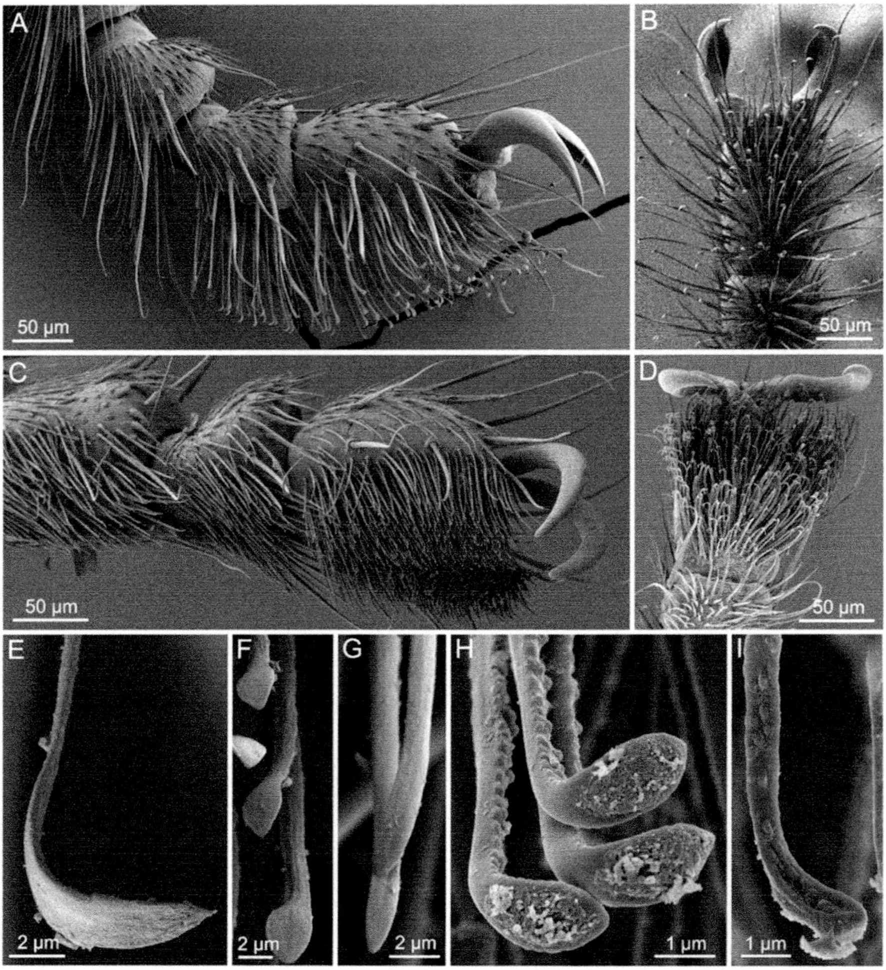

Fig. 3.1 Scopula in Laniatores. Distal tarsi of hind legs. (**a–b, e**) Epedanidae: *Metacrobunus* sp. Spatulate setae are present in the two distal most tarsomeres. The brush of these modified setae is called scopula. (**c–d, f–i**) Biantidae. The scopula is only present on the distal most segment, which is slightly broadened. The spatulate setae are much shorter, have smaller tips and stand much denser than those in *Metacrobunus* sp.. (**c, f–g**) *Hinzuanius flaviventris* (Biantinae); (**d, h–i**) *Galibrotus* cf. *riedeli* (Stenostygninae)

Like in insects, a fluid supplement of the scopula of Laniatores is likely. In *G. cf. riedeli,* there were often clumps of an amorphous substance on the ventral side of the spatulate tips (Fig. 3.1h, i), which might be remains of a coagulated secretion. How the secretion is delivered so locally is unclear, because no pores or channels were discernable in SEM images.

In the cryptic hooded tickspiders (Ricinulei) the structure of the scopula is rather similar. The scopula is present on the flattened ventral sides of the distal tarsal segment of the walking legs (Fig. 3.2a, b), with the exception of the elongated second

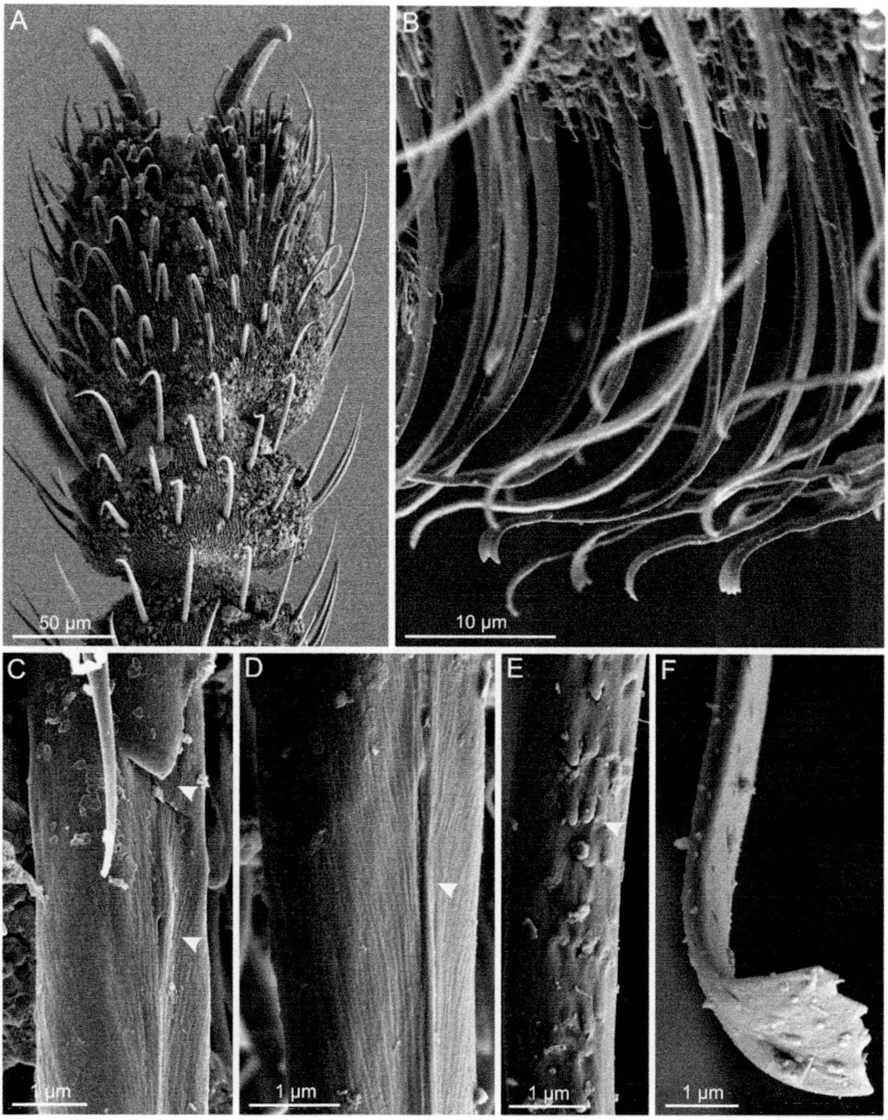

Fig. 3.2 Scopula of Ricinulei. *Pseudocellus pachysoma* (Ricinoididae), nymph. (**a**) Tarsus of leg 4, ventral view. Spatulate setae are present only on the distal most tarsomere. (**b**) Detail of scopula, lateral view. (**c**–**f**) Details of spatulate setae. (**c**) Basis with slit-like channel openings (*arrowheads*). (**d**) Setal shaft with groove (*arrowhead*), presumably supporting secretion transport. (**e**) Setal back, with nubs (*arrowhead*). (**f**) Apical part with spatulate tip

leg pair, which is used as feelers. The spatulate setae exhibit a rear fold (Fig. 3.2c, d), which stands in connection with secretion channels (Talarico et al. 2006). The rear side of both shaft and spatulae are irregularly covered by small nubs (Fig. 3.2e), similar to the scopula setae of some harvestmen (see above) or rear sides of spatulae in some bugs (Weihrauch, 2007). The similarity of the structure to hairy adhesive pads of insects, the location on the tarsal sole of the walking legs and the presence of secretion channels suggest an adhesive function. However, this was not experimentally tested yet. The ecological benefit of an adhesive pad is quite unclear in this case. The species studied here (*P. pachysoma*) lives deep inside a cave, where the animals were exclusively found under stones buried in a thick layer of guano (Teruel and Schramm 2014). A need of climbing seems to be ambiguous under these circumstances. It is unknown, if the scopula is also present in ricinuleid genera other than *Pseudocellus*.

3.3 Spatulate Microtrichia

Much more common than 'simple' tenent setae with a single spatulate tip are setae with spatulate microtrichia. The setal section, which bears the microtrichia, is often spatula-shaped itself: It is broadened, flattened and therefore highly flexible. This means, these setae exhibit a bi-hierarchical spatula structure. However, there are also tenent setae that are tapered like usual setae, multiple branched or even spine-like. In a single case spatulate microtrichia have also been reported from urticating hairs, which are discarded by tarantulas to deter predators (Perafán and Pérez-Miles 2010). In this case the structures may assist the attachment to the predator skin. In some mites, similar spatulate microtrichia, as those on setae, may occur on the claws or the empodium.

3.3.1 Tenent Setae of Spiders

Tenent setae evolved for multiple times among spiders (Wolff et al. 2013; Ramírez 2014). They consist of a cylindrical shaft and a usually highly flattened and broadened tip, which is densely covered with spatula bearing microtrichia on one side (Homann 1957; Foelix and Chu-Wang 1975; Hill 1977) (Figs. 3.3a and 3.5). The spatulae are triangular shaped in araneomorphs (Fig. 3.3d), circular in mygalomorphs (Fig. 3.3b), and oval in the scopulae (but not claw tufts) of Palpimanidae (Fig. 3.3c) (Wolff et al. 2013). The elaborate microstructure of the tenent setae was found to be responsible for remarkable climbing abilities, even on very smooth surfaces, like glass slides. Homann (1957) hypothesized that their adhesion is based on the interaction with adsorbed water on the substrate surface. Since then, different hypothesis were raised regarding the function of the setae, including electrostatic forces (Homann 1957), suction (Hill 1977), interlocking with minute asperities (Hill 1977), van-der-Waals forces (Kesel et al. 2003) and capillary forces of a fluid secretion (Peattie et al. 2011). Adhesion (Kesel et al. 2003) and friction measurements with fresh and dried samples (Niederegger and Gorb 2006; Wolff and Gorb 2013) revealed that the tenent setae are basically structural, dry adhesives, but adsorbed water may enhance adhesion (Wolff and Gorb 2012b). Because only one side of the flexible distal setal section is covered with the spatulate microtrichia, friction is highly anisotropic (Niederegger and Gorb 2006; Wolff and Gorb 2013). When sliding in one direction, the microtrichia of the setal lamella align and get in

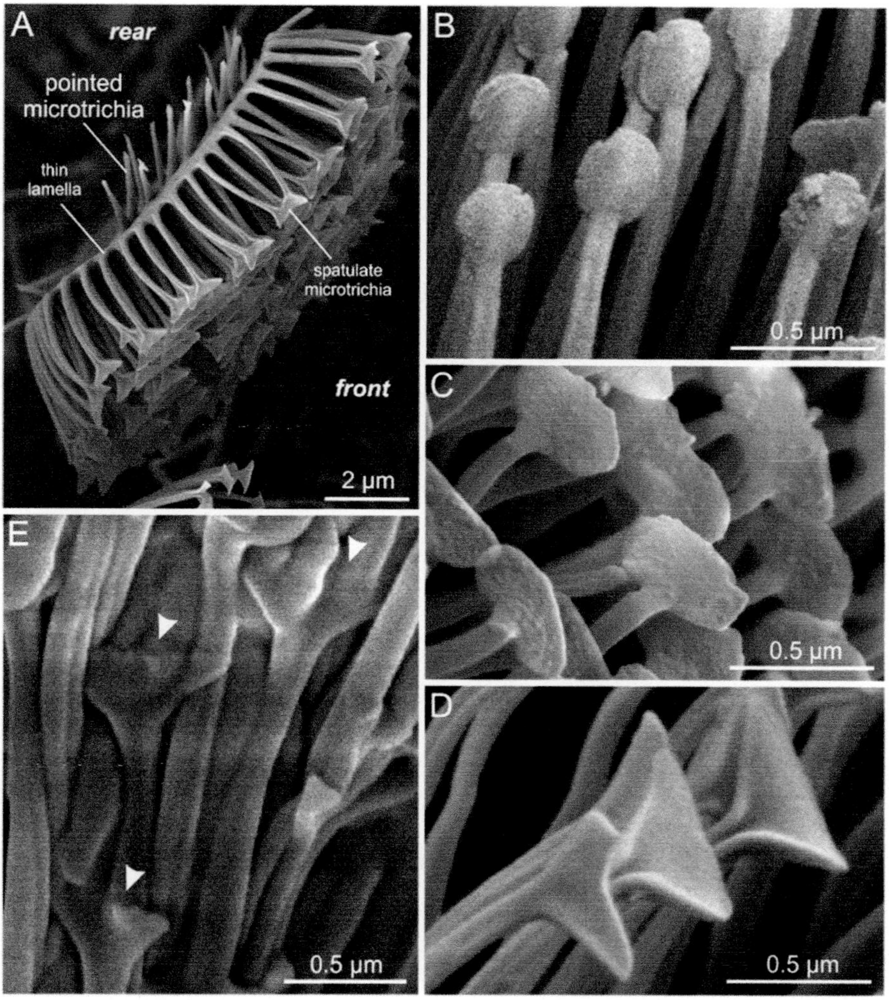

Fig. 3.3 Tenent setae of spiders. (**a**) The spatulate microtrichia are arranged only on one side of the flattened setal tip, like seen here in a claw tuft seta of the jumping spider *Euophrys frontalis* (apical view). The rear side bears rows of short pointed microtrichia that are assumed to be anti-sticking structures. Note, that at this apical part the seta builds a lamella-like structure that is only about 100 nm thick and therefore highly flexible. (**b–d**) The spatula shape differs between different lineages of spiders that have evolved the tenent setae independently: (**b**) circular spatula in tarantulas (here: *Aphonopelma seemanni*; image by J. Kottsieper), (**c**) oval spatulae in the scopulae (but not claw tufts) of Palpimanidae (here: *Palpimanis gibbulus*), (**d**) triangular spatulae as typical for Araneomorphae (here: *E. frontalis*), and analogously in trombidiform mites (see Fig. 3.7). (**e**) Spatular dimensions and shapes presumably arise from the fibrillar structure of the arachnid cuticle. The formation of spatulae (*arrowheads*) can be seen in this premature seta, which was dissected from a pre-molt *Anyphaena accentuata*

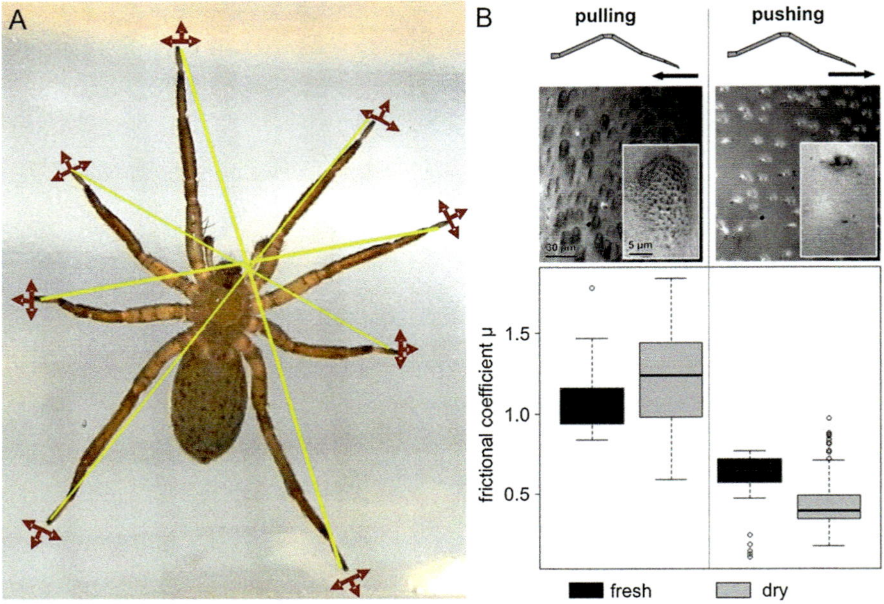

Fig. 3.4 Tribology of the hairy adhesive pads. Due to the polar arrangement of their spatulate microtrichia the tenent setae of spiders show highly anisotropic friction. The orientation of setae differs in different parts of the claw tuft, which can be brought in contact in variable proportions. (**a**) This image illustrates the directions of the highest friction, and the cooperative action of contra-lateral legs in generating a strong attachment (photo: *Zora spinimana* on a vertical glass slide; photo by S. Voss, with kind permission). (**b**) The amount of friction corresponds to the contact area, which was visualised by reflection interference contrast microscopy (second row of images shows detail of claw tuft setae, insets show contact of a single seta, dark spots indicate real contact between the spider structures and substrate). Drying does not reduce but even enhances the effect, supporting the hypotheses that it is based on dry adhesion. Data were obtained on the ctenid spider *Cupienniussalei*; (Images are adapted from Wolff and Gorb 2013)

contact with the substrate, which results in high friction (Fig. 3.4b left column). When sliding in the opposite direction the setal lamella turns over due to its flexibility, the spatular contacts break and friction drops (Fig. 3.4b right column) (Video 3.8). This permits both the generation of high attachment forces and in the next moment a very quick and effortless detachment. In large spiders, the setae are differently orientated in different parts of the pad, and depending on which part is used, the frictional anisotropy works in different directions (Wolff and Gorb 2013) (Fig. 3.4). This mechanism permits a precise control of the attachment, with the pads of contra-lateral legs working synergistically (Wohlfart et al. 2014).

The adhesion of the tenent setae is highest on very smooth surfaces and lowest at surfaces with roughness at the micron scale, where the asperity size is similar to the width of the spatulae (Wolff and Gorb 2012c). The performance of the adhesive pads is also highly influenced by humidity. It increases with relative humidity and drops at water saturation of the air followed by condensation, which leads to aquaplaning effects (Wolff and Gorb 2012b).

The tenent setae of spiders are hypothesized being evolved from frictional setae (see Chap. 2), which are flattened and uni-directionally covered with microtrichia, too, but lack both spatulae and a broadening of the setal tip (Wolff and Gorb 2012a; Wolff et al. 2013). Their evolution is highly correlated with the abandoning of building prey capture webs (Wolff et al. 2013). Tenent setae obviously represent a considerable ecological benefit, because the majority of free hunting spiders is equipped with them. Due to their arrangement on the legs in, so-called, *scopulae*, they play an important role in prey capture, and as *claw tufts* at the tips of the legs, they support locomotion on smooth substrates.

Scopula Tenent setae often occur in thin strips or dense brushes at the pro- and retrolateral sides of the distal leg segments (Fig. 3.6a, b). These, so-called, scopulae primarily assist prey capture (Rovner 1978; Foelix et al. 1984; Pekár et al. 2011; Eggs et al. 2015). In large spiders, they may also substantially support the claw tufts during locomotion (Pérez-Miles et al. 2015).

The scopula setae are connected to the underlying cuticle through oval sockets that permit a directed movement. In rest, they lay flat on the leg and similar to the spines (see Chap. 2) can get erected to a certain extent by the increased hemolymph pressure during an attack (Rovner 1978; Rovner 1980; Eggs et al. 2015). The erectable spines are important for retaining the captured prey (see Chap. 2). In comparison to the spines, which are usually lacking on the tarsus, the scopulae are often largest in the tarsi, which may suggest a different function, namely the seizing of the prey (Eggs et al. 2015).

In spiders that use the front legs during attack (e.g. Gnaphosidae), the scopula is restricted to these legs, while in others, which use a full capture basket (e.g. Philodromidae), the scopula is equally evolved in all legs (Wolff and Gorb 2012a; Eggs et al. 2015). The Palpimanidae only use the first pair of legs, which is highly modified in this family, lacking claws and spines, and hence totally relying on the adhesive capacity of the scopula (Pekár et al. 2011) (see image of the palpimanid front leg in Chap. 2, and Video 2.8).

Claw Tuft The claw tufts are dense brushes of tenent setae at the distal tip of the tarsus (Fig. 3.6). They primarily serve friction generation and attachment on smooth and micro rough surfaces during locomotion. The claw tuft setae show some modifications if compared to the scopula. They are twistable and bendable in different directions (Fig. 3.5). This presumably lowers the stress acting on the bonding and stabilizes the contact during the foot rotation, when the spider pushes its body forwards (Wolff et al. 2013). The basal shaft can be highly elongated, especially in large spiders and especially in the distal part of the pad. It often exhibits an S-like shape, which may work like a spring in order to absorb mechanical stress (Gasparetto et al. 2009).

The claw tuft setae are situated on a sclerite, the tenent plate, which is partly or totally separated from the tarsus by flexible, non-sclerotized cuticle (Hill 2006) (Fig. 3.6a). The sockets are packed very densely, to provide space for as much tenent setae as possible. This is because in spiders normally only the distal tarsal tip and the pretarsal structures contact the ground during locomotion. A considerable

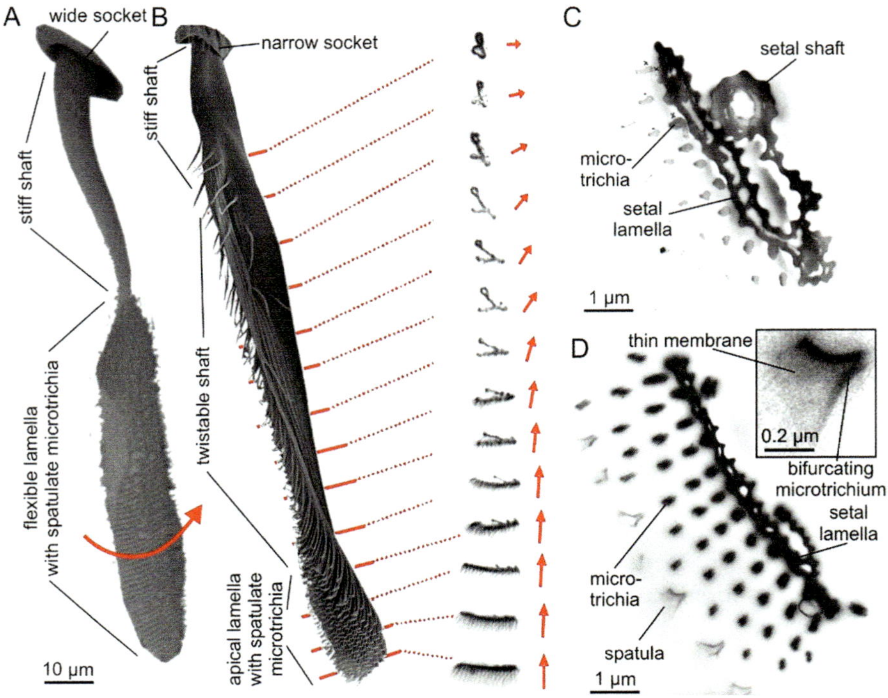

Fig. 3.5 **Functional difference between scopula and claw tuft setae**. (**a**) The scopula seta exhibits a stiff, cylindrical shaft and a clearly delimitated apical lamella, which is bendable in one main direction (*red arrow*). (**b**) In contrast, the claw tuft seta exhibits a twisted lamella (see series of TEM sections in the *right*) and a gradient of the flexibility, as illustrated in the image: the length of the *arrows* indicates the degree of flexibility, their direction corresponds to the angle of deflection. (**c**–**d**) Details of the fine structure in a claw tuft seta of *Selenops mexicanus*, with (**c**) a section from the proximal part of the lamella (with attached shaft), and (**d**) a section from the distal part of the lamella (with spatulate microtrichia); *inset* shows detail of a spatula (Images by W. Nentwig, with kind permission)

pad area is achieved by the distal broadening of the tenent setae and an increase of the length of setal shafts from proximal to distal. This results in a pad area that is much higher than the tenent plate area. In rare cases there are only one or two single enlarged tenent setae beneath the claws, like in *Goyenia* (Desidae) and *Apostenus* (Liocranidae) (Ubick and Vetter 2005). The tenent plates may be indirectly moveable by a mechanical coupling with the pretarsus or changes in hemolymph pressure (Dunlop 1994; Hill 2006; Wolff et al. 2013) (Fig. 3.6c, d). In other instances there are interlocking structures or hooks ('claw tuft claspers') on the basis of the claws that are assumed to manipulate the position and angle of the setae (Ramírez 2014) (Fig. 3.6e, f).

Claw tufts convergently evolved in mygalomorphs, haplogynes and for multiple times in entelegyne spiders (Wolff et al. 2013). They are nearly restricted to hunting spiders with the exception of the small family Psechridae, with web-building species that presumably evolved from free hunting ones. Hunting spiders with claw

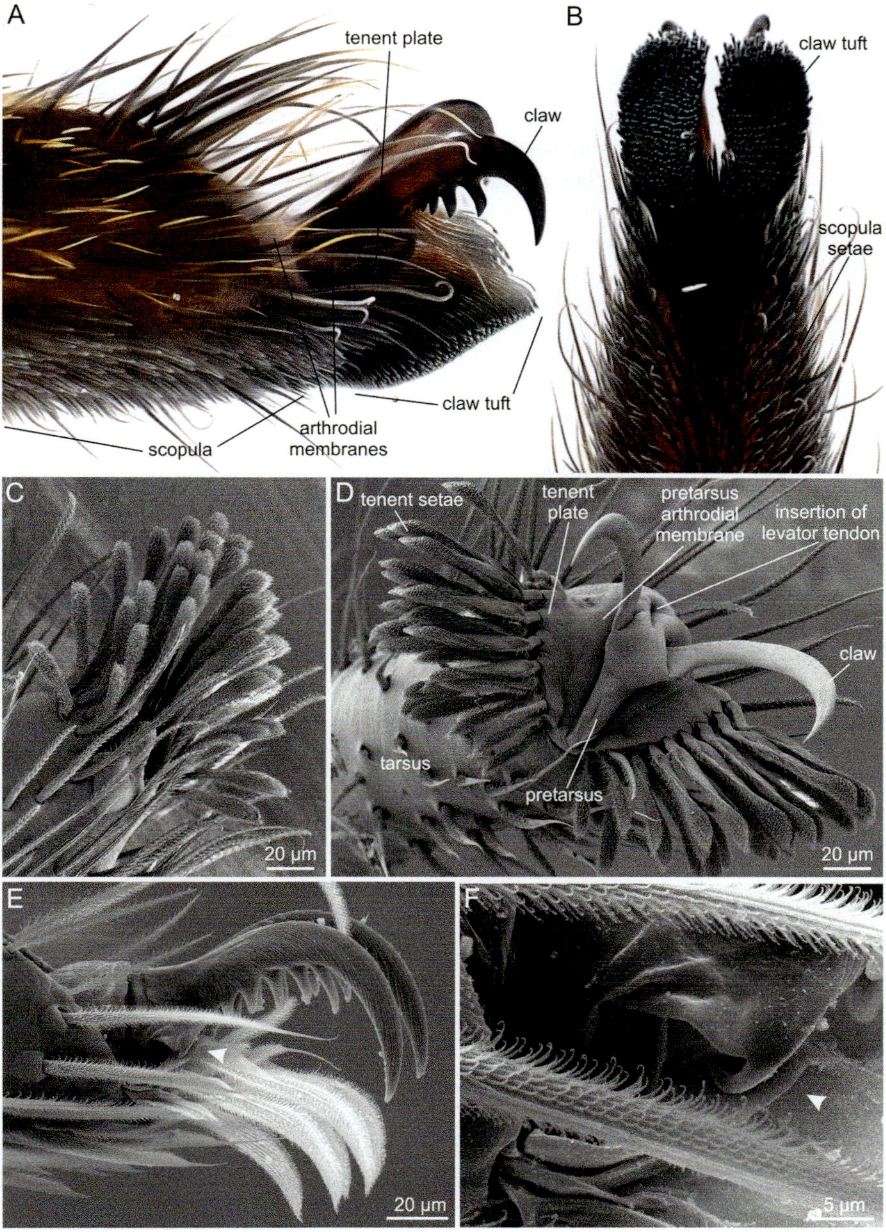

Fig. 3.6 Movability of the claw tuft. The claw tuft is a special, partitioned modification of the distal tarsus, forming a stiff, slerotized plate (tenent plate) of densely packed sockets, in which the tenent setae are based. (**a**) The tenent plate is often delimitated by membranous areas, as in the large hunting spider *Phoneutria fera*. Hence it follows the movements of the pretarsus, when either the protractor or retractor muscle contracts (image by B. Eggs, with kind permission). (**b**) The claw tufts are always paired, but usually appear as a single pad. (**c–d**) Increased hemolymph pressure leads to separation and lateral spreading of the paired pads, as simulated here by slightly squeezing a leg of the jumping spider *Euophrys frontalis* and visualizing it in the shock frozen state in the Cryo-SEM (adapted from Wolff et al. 2013). (**e–f**) Spiders of a particular dionychan clade (here an un-determined species of Liocranidae) exhibit modified hooks, so-called 'claw-tuft-claspers' at the base of the claws. These interlock with a fold at the base of the tenent setae (*arrowhead* – note, that in this preparation the clasper is de-coupled), causing a mechanical coupling for a simultaneous movement of the setae with the pretarsus (Images by M. Ramírez, with kind permission)

tufts are significantly more abundant in the higher vegetation strata than at the ground and herbal stratum, whereas the possession of claw tufts did not as obviously correlate with the microhabitat specialization of species (Wolff and Gorb 2014). This indicates that claw tuft bearing species are ecologically more successful in above-ground habitats and less competitive at the ground. The benefit should be primarily due to a higher safety factor and increased maneuverability than an advantage in the attachment ability, because the demand of climbing might also be high in highly structured ground habitats, like deciduous leaf litter.

3.3.2 Tenent Microtrichia of Trombidiform Mites

Spatulate microtrichia (often erroneously called 'tenent setae' in the literature) are widespread among Trombidiformes, but are absent from the Sarcoptiformes and all parasitiform orders. These microtrichia may be located on special branched tarsal setae (Anystidae: Anystinae, and Trombidiidae: *Allothrombium*), modified claws (some Tetranychoidea) (Fig. 3.7b), or the (then often broadened) empodium (some Tetranychoidea, Cheyletoidea, and Raphignathoidea, Bdelloidea, some Eupodoidea, Tydeoidea and Eriophyoidea) (Fig. 3.7d–g) (Alberti and Coons 1999; Krantz and Walter 2009; Voigt 2016). The tenent empodium seems to be the plesiomorphic state, since it is found in all sub-lineages (in Eupodina and Eleutherengona widespread, in Anystina to our knowledge only in larvae of Erythraeidae, and in the basal Sphaerolichida among Lordalichidae (Krantz and Walter 2009) but unclear if microtrichia are spatulate in the latter). Basically, the structure of the spatulate microtrichia is strikingly similar to those of the tenent setae of araneomorph spiders: the shaft consists of two fibrils that apically bifurcate, spanning up a thin membrane between them (Fig. 3.7j). This striking analogy is presumably based on a similar precursor structure, namely a specific fibrillary structure of the arachnid cuticle. In gall mites (Eriophyoidea) a variety of tip shapes of tenent microtrichia has evolved: spatula-, band- or disc-like (Krantz and Walter 2009).

The Erythraeidae possess branched setae in the ventral tarsi exhibiting band-like microtrichia that lack spatula (Fig. 3.7h) (Alberti and Coons 1999). We studied these setae in *Abrolophus* sp. and found that they are only present in nymphs and adults (Fig. 3.7c). The three-legged larva lacks them, but bears an empodium with

Fig. 3.7 (continued) which is an artefact). (**c**) In the Erythraeidae (here: *Abrolophus* sp., leg 1) adults and nymphs bear small tenent setae lacking spatulae (note the absence of the empodium, *arrowhead*). (**d**) In contrast, the larva (here: same species as in (**c**), leg 1) lacks the setae and bears a tenent empodium with spatulate microtrichia. (**e–f**) In the Cocceupodidae (here: *Linopodes* sp., (**e**) leg 1, (**f**) leg 4) and (**g**) the snout mites (Bdellidae, here: *Cyta* sp., leg 1) a tenent empodium is present in all walking legs. (**h–j**) The fine structure of tenent microtrichia differs: (**h**) In the tenent setae of Erythraeidae (here: *Abrolophus* sp.) the tips of microtrichia are tape like and lack spatulae; whereas (**i**) in snout mites (here: *Cyta* sp.) and (**j**) spider mites (here: *Bryobia* sp.) the tips bear triangular spatulae. Note the extreme similarity with the analogue tenent microtrichia of spiders (Fig. 3.3d)

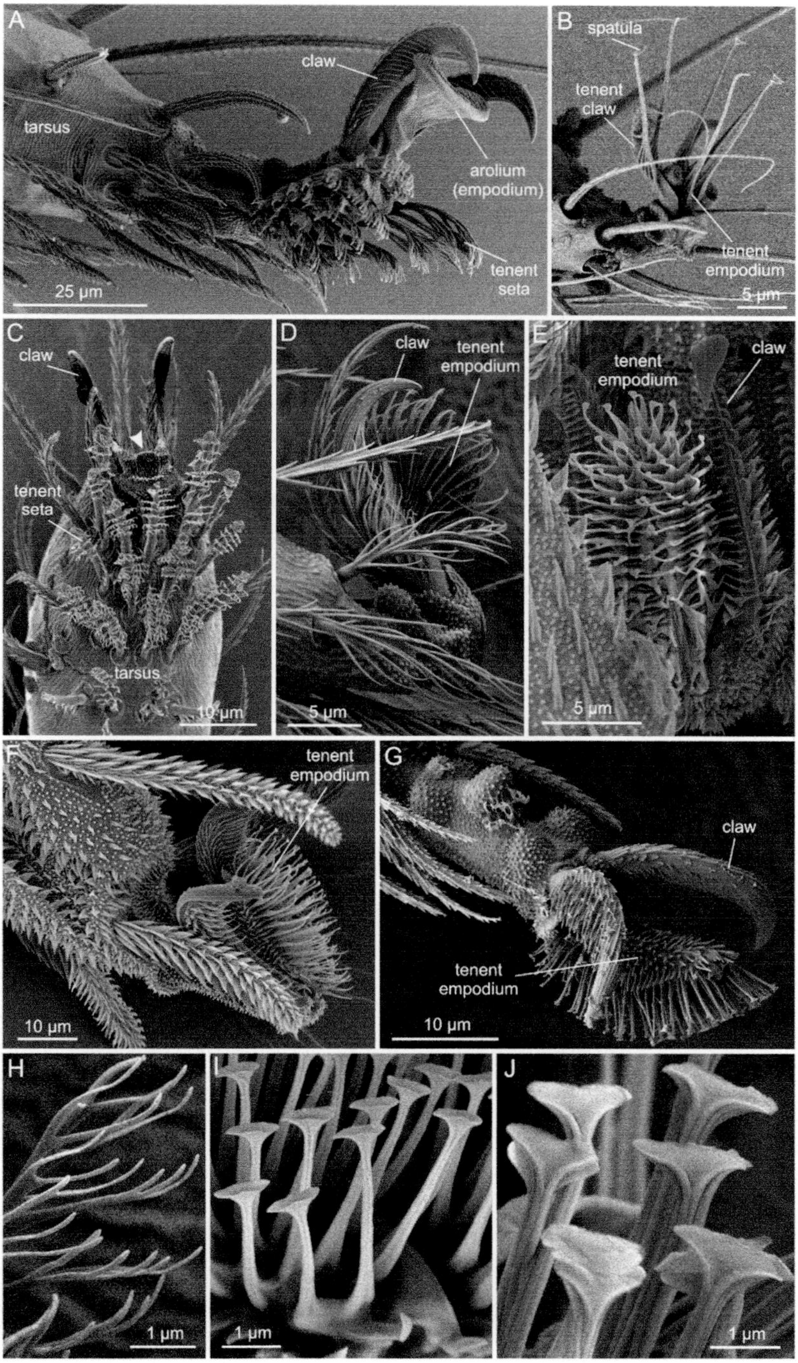

Fig. 3.7 Tenent microtrichia in mites (Trombidiformes). Cryo-SEM images. (**a**) Mites of the subfamily Anystinae (here: *Anystis* sp., leg 2) bear paired large, multiple branched tenent setae with spatulate microtrichia. The empodium is modified into an arolium recruiting additional adhesive forces (see Chap. 7). (**b**) In spider mites (here: *Bryobia* sp., leg 1) the tenent microtrichia are located on the claws and the empodium (2 parallel rows each; in this sample the spatulae are stuck together,

spatulate microtrichia instead (Fig. 3.7d). Both larvae and adults are capable to scale smooth surfaces with high speed regardless of the differences in the structure of tenent microtrichia.

The fast running *Anystis* spp. (Anystidae) possess a pair of large multiple branched tenent setae, but also an arolium in each of the eight walking legs (Fig. 3.7a). This is a unique configuration among arachnids. We found that usually both the arolium and the tenent setae are brought into contact during locomotion on smooth surfaces (see Chap. 7; Video 7.19, but see Video 7.21 where only the setae are brought in contact). It remains unclear in which situation which of the pads (hairy or smooth) are dominating the attachment function.

Lipidoid fluid footprints were found in the tenent setae of *Tetranychus urticae* (Tetranychidae) (Mizutani et al. 2006), *Balaustium murorum* (Erythraeidae) (Peattie et al. 2011), *Abrolophus* sp. (Erythraeidae) and *Anystis* sp. (this study, see Chap. 7). Hence, capillary forces may play a role in the adhesive function of these setae. However, theoretical models and measurements on substrates with different free surface energy showed that at this small scale the intermolecular attractive forces between spatula and substrate play the dominant role (Mizutani et al. 2006). A study investigating shock frozen spider mites on various plant surfaces did not reveal any fluids in the contact (Voigt 2016). The adhesive forces were measured in spider mites (*Tetranychus urticae*), which bear four spatulate microtrichia in each ambulacrum (Mizutani et al. 2006). The adhesion is about 8 μN on glass, which corresponds to a safety factor of 37 (Mizutani et al. 2006). The adhesion is reduced by high humidity and surface roughness (Mizutani et al. 2006). The tenent setae are the only attachment devices on the legs of these animals and hence play the dominant role in attachment and locomotion on the inhabited leaf surfaces (Voigt 2016).

3.4 Spatulate Surface Structures

The arolium of whip-spiders (Amblypygi) has a unique structure, combining features of smooth and hairy adhesive pads (see also Chap. 7). The contact is made by a distal transverse-oval lip, in which the cuticle is bloated and consists of numerous parallel fibers that are branching multiple times towards the epicuticle (Wolff et al. 2015). In the distal part the fine branches are merging again, forming hollow hexagonal tubes, which are apically keeled and equipped with a spatula-like tip (Fig. 3.8c). This specific microstructure leads to a discontinuous peel-off behavior (Video 7.11), which highly enhances the adhesive strength (see Chap. 10). This kind of contact behavior has never been observed in any other arolium before. The amblypygid arolium is sticky both with and without apparent secretions (Videos 7.11, 7.12, 7.13 and 7.14), and can therefore be considered as a facultative dry adhesive. Spatulae are not yet developed in the arolium of the first instar, the pre-nymph, which settles on its mother.

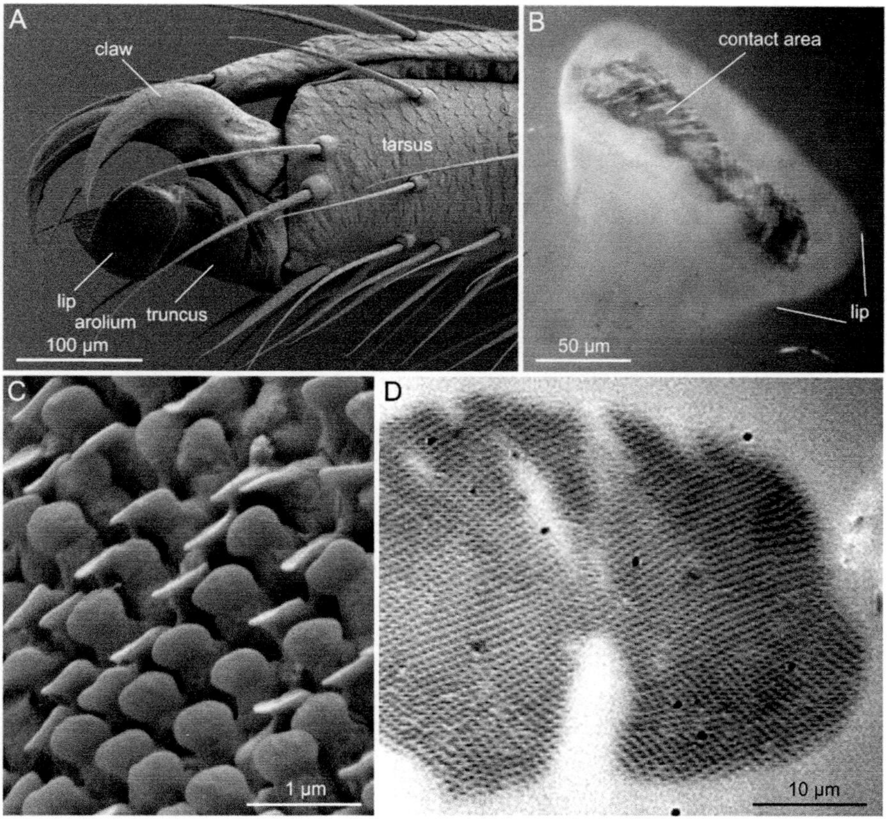

Fig. 3.8 Spatulae of the whip-spider arolium. (**a**) Whip-spiders of the families Charinidae and Charontidae (here: *Charon* cf. *grayi*, protonymph, leg 2) exhibit an arolium with an apical, flexible lip that generates the contact with the substrate. (**b**) The contact area can be visualized with coaxial illumination. (**c**) The lip bears spatulate structures of comparable size and shape than that of tenent microtrichia. (**d**) Reflection interference microscopy reveals details of the actual contact area, with only the spatulae being in contact

3.5 Micro-tapes

In the Araneomorphae the silk of the major ampullate glands is the most important and the toughest fibre (Blackledge and Hayashi 2006). It exhibits a thin coat of glycoproteins (Sponner et al. 2007) that may produce some adhesion with substrates. However, the adhesion is usually weak, because of a small contact area, given by the cylindrical shape and comparably low compliance (Wolff et al. 2014). Brown recluse spiders (Sicariidae: *Loxosceles*) have evolved modified spigots (gland openings) that are slit-like and consequently extrude ribbon-like fibres (Knight and Vollrath 2002). This kind of fibres is highly compliable and can generate a high contact area (Fig. 3.9). They are barely removable from a smooth glass although there is

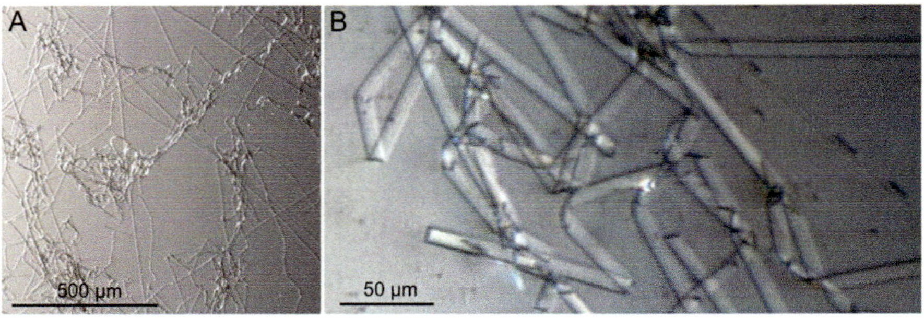

Fig. 3.9 Tape-like major ampullate gland silk of *Loxosceles rufescens*. Transmission light micrographs

obviously no glue involved. The mechanism is probably based on dry adhesion and can be compared with the contact mechanics of thin elastic polymer films.

Other micro-tapes are represented by some rare cases of tenent setae or microtrichia, with the distal end flattened but not significantly broadened. This applies to the setae of some parts of the scopula of Laniatores harvestmen (Rambla 1990) (Fig. 3.1g), and the microtrichia of tenent setae in mites of the family Erythraeidae (Fig. 3.7h) and some gall mites (Eriophyoidea) (Alberti and Coons 1999). In hairy adhesive pads, such structures might be disadvantageous, if compared with spatulae due to the reduced peeling line. However, we found that the erythraeid *Abrolophus* sp. could easily scale smooth surfaces, although it had a comparably small contact area (see above).

References

Alberti G, Coons L (1999) Acari: Mites. In: Harrison FW, Foelix RF (eds) Microscopic anatomy of invertebrates: Chelicerate Arthropoda, vol 8. Wiley-Liss, New York, pp 515–1265

Arzt E, Gorb S, Spolenak R (2003) From micro to nano contacts in biological attachment devices. Proc Natl Acad Sci 100(19):10603–10606

Autumn K, Dittmore A, Santos D, Spenko M, Cutkosky M (2006) Frictional adhesion: a new angle on gecko attachment. J Exp Biol 209(18):3569–3579

Blackledge TA, Hayashi CY (2006) Silken toolkits: biomechanics of silk fibers spun by the orb web spider *Argiope argentata* (Fabricius 1775). J Exp Biol 209(13):2452–2461. doi: 10.1242/Jeb.02275

Chung JY, Chaudhury MK (2005) Roles of discontinuities in bio-inspired adhesive pads. J R Soc Interface 2(2):55–61

Dunlop JA (1994) Movements of scopulate claw tufts at the tarsus tip of a tarantula spider. Neth J Zool 45(3):513–520

Eggs B, Wolff JO, Kuhn-Nentwig L, Gorb SN, Nentwig W (2015) Hunting without a web: how lycosoid spiders subdue their prey. Ethology 121(12):1166–1177

Federle W (2006) Why are so many adhesive pads hairy? J Exp Biol 209(14):2611–2621

Filippov A, Popov VL, Gorb SN (2011) Shear induced adhesion: contact mechanics of biological spatula-like attachment devices. J Theor Biol 276(1):126–131

Foelix RF, Chu-Wang I-W (1975) The structure of scopula hairs in spiders. In: Proceedings of the 6th international arachnological Congress. Amsterdam, Nederlandse Entomologische Vereniging, pp 156–157

Foelix RF, Jackson RR, Henksmeyer A, Hallas S (1984) Tarsal hairs specialized for prey capture in the salticid *Portia*. Rev Arachnol 5:329–334

Gasparetto A, Seidl T, Vidoni R (2009) A mechanical model for the adhesion of spiders to nominally flat surfaces. J Bionic Eng 6(2):135–142

Gorb SN (1998) The design of the fly adhesive pad: distal tenent setae are adapted to the delivery of an adhesive secretion. Proc R Soc B 265(1398):747–752

Gorb S (2001) Attachment devices of insect cuticle. Springer Science & Business Media, Dordrecht/Boston/London, 305pp

Gorb S, Beutel R (2001) Evolution of locomotory attachment pads of hexapods. Naturwissenschaften 88(12):530–534

Haas F, Gorb S (2004) Evolution of locomotory attachment pads in the Dermaptera (Insecta). Arthropod Struct Dev 33(1):45–66

Hill DE (1977) The pretarsus of salticid spiders. Zool J Linn Soc Lond 60(4):319–338

Hill DE (2006) Jumping spider feet (Araneae, Salticidae). Peckhamia Epublications 3:1–41

Homann H (1957) Haften Spinnen an Einer Wasserhaut. Naturwissenschaften 44(11):318–319

Jagota A, Bennison SJ (2002) Mechanics of adhesion through a fibrillar microstructure. Integr Comp Biol 42(6):1140–1145

Kendall K (1975) Thin-film peeling – elastic term. J Phys D Appl Phys 8(13):1449–1452

Kesel A, Martin A, Seidl T (2003) Adhesion measurements on the attachment devices of the jumping spider *Evarcha arcuata*. J Exp Biol 206(16):2733–2738

Knight DP, Vollrath F (2002) Spinning an elastic ribbon of spider silk. Phil Trans R Soc B 357(1418):219–227

Krantz GW, Walter DE (2009) A manual of acarology, 3rd edn. Texas Tech University Press, Lubbock, 816pp

Labonte D, Clemente CJ, Dittrich A, Kuo C-Y, Crosby AJ, Irschick DJ, Federle W (2016) Extreme positive allometry of animal adhesive pads and the size limits of adhesion-based climbing. Proc Natl Acad Sci 113(5):1297–1302

Mizutani K, Egashira K, Toukai T, Ogushi J (2006) Adhesive force of a spider mite, *Tetranychus urticae*, to a flat smooth surface. JSME Int J Ser C 49(2):539–544

Niederegger S, Gorb SN (2006) Friction and adhesion in the tarsal and metatarsal scopulae of spiders. J Comp Physiol A 192(11):1223–1232

Niederegger S, Gorb S, Jiao Y (2002) Contact behaviour of tenent setae in attachment pads of the blowfly *Calliphora vicina* (Diptera, Calliphoridae). J Comp Physiol A 187(12):961–970

Peattie A, Full R (2007) Phylogenetic analysis of the scaling of wet and dry biological fibrillar adhesives. Proc Natl Acad Sci 104(47):18595–18600

Peattie AM, Dirks J-H, Henriques S, Federle W (2011) Arachnids secrete a fluid over their adhesive pads. Plos One 6(5):e20485

Pekár S, Šobotník J, Lubin Y (2011) Armoured spiderman: morphological and behavioural adaptations of a specialised araneophagous predator (Araneae: Palpimanidae). Naturwissenschaften 98(7):593–603

Perafán C, Pérez-Miles F (2010) An unusual setule on type IV urticating setae of Homoeomma uruguayense (Araneae: Theraphosidae). J Arachnol 38(1):153–154

Peressadko A, Gorb SN (2004) When less is more: experimental evidence for tenacity enhancement by division of contact area. J Adhes 80(4):247–261

Pérez-Miles F, Perafán C, Santamaría L (2015) Tarantulas (Araneae: Theraphosidae) use different adhesive pads complementarily during climbing on smooth surfaces: experimental approach in eight arboreal and burrower species. Biology Open: bio.013144

Pinto-da-Rocha R, Machado G, Giribet G (2007) Harvestmen: the biology of Opiliones. Harvard University Press, Cambridge, MA, p 608

Qian J, Gao H (2006) Scaling effects of wet adhesion in biological attachment systems. Acta Biomat 2(1):51–58

Rambla M (1990) Les scopula des Opilions, differences avec les scopula des Araignées (Arachnida, Opiliones, Araneae). Bull Soc Europ Arachnol 1:293–298

Ramírez MJ (2014) The morphology and phylogeny of dionychan spiders (Araneae, Araneomorphae). Bull Amer Mus Nat Hist 390:1–374

Roewer CF (1923) Die Weberknechte der Erde. Gustav Fischer, Jena, p 1116

Rovner JS (1978) Adhesive hairs in spiders: behavioral functions and hydraulically mediated movement. Symp Zool Soc Lond 42:99–108

Rovner JS (1980) Morphological and ethological adaptations for prey capture in wolf spiders (Araneae, Lycosidae). J Arachnol 8:201–215

Spolenak R, Gorb S, Arzt E (2005) Adhesion design maps for bio-inspired attachment systems. Acta Biomater 1(1):5–13

Sponner A, Vater W, Monajembashi S, Unger E, Grosse F, Weisshart K (2007) Composition and hierarchical organisation of a spider silk. Plos One 2(10):e998

Talarico G, Palacios-Vargas JG, Fuentes Silva M, Alberti G (2006) Ultrastructure of tarsal sensilla and other integument structures of two *Pseudocellus* species (Ricinulei, Arachnida). J Morphol 267(4):441–463

Teruel R, Schramm FD (2014) Description of the adult male of *Pseudocellus pachysoma* Teruel & Armas 2008 (Ricinulei: Ricinoididae). Rev Ibér Aracnología 24:75–79

Ubick D, Vetter RS (2005) A new species of *Apostenus* from California, with notes on the genus (Araneae, Liocranidae). J Arachnol 33(1):63–75

Varenberg M, Pugno NM, Gorb SN (2010) Spatulate structures in biological fibrillar adhesion. Soft Matter 6(14):3269–3272

Voigt D (2016) In situ visualization of spider mite–plant interfaces. J Acarol Soc Jpn 25(S1):119–132

Weirauch C (2007) Hairy attachment structures in Reduviidae (Cimicomorpha, Heteroptera), with observations on the fossula spongiosa in some other Cimicomorpha. Zool Anz 246(3):155–175

Wohlfart E, Wolff JO, Arzt E, Gorb SN (2014) The whole is more than the sum of all its parts: collective effect of spider attachment organs. J Exp Biol 217(2):222–224

Wolff JO, Gorb SN (2012a) Comparative morphology of pretarsal scopulae in eleven spider families. Arthropod Struct Dev 41(5):419–433

Wolff JO, Gorb SN (2012b) The influence of humidity on the attachment ability of the spider *Philodromus dispar* (Araneae, Philodromidae). Proc R Soc B 279(1726):139–143

Wolff JO, Gorb SN (2012c) Surface roughness effects on attachment ability of the spider *Philodromus dispar* (Araneae, Philodromidae). J Exp Biol 215(1):179–184

Wolff JO, Gorb SN (2013) Radial arrangement of Janus-like setae permits friction control in spiders. Sci Rep 3:1101

Wolff JO, Gorb SN (2014) Adhesive foot pads: an adaptation to climbing? An ecological survey in hunting spiders. Zoology 118:1–7

Wolff JO, Nentwig W, Gorb SN (2013) The great silk alternative: multiple co-evolution of web loss and sticky hairs in spiders. Plos One 8(5):e62682

Wolff JO, Schneider JM, Gorb SN (2014) How to pass the gap – Functional morphology and biomechanics of spider bridging threads. In: Asakura T, Miller T (eds) Biotechnology of silk, vol 5. Springer, Dordrecht, pp 165–177

Wolff JO, Seiter M, Gorb SN (2015) Functional anatomy of the pretarsus in whip spiders (Arachnida, Amblypygi). Arthropod Struct Dev 44(6):524–540

Chapter 4
Nano-fibres

Abstract High adhesion demands a large area of close (real) contact, which can be achieved by film-, spatula-, tape- or mushroom-like structures. However, at sub-micron scale, very high aspect ratio fibres, exhibiting a cylindrical shape may get tacky due to their high surface-to-volume ratio. Bundles of delicate, tough nano-fibres may recruit dry adhesion by the action of van der Waals, hygroscopic or electrostatic forces, as demonstrated for the cribellate capture threads and bridging threads of spiders. Additionally, there are other adhesive structures based on nano-fibres, such as, for example, special microtrichous setae of some spiders camouflaging themselves with sand particles, and the hairy soles of harvestmen.

As discussed in the previous chapter, compressed tape-like structures can generate strong adhesion by creating a large area of close contact with the substrate. A cylindrical structure, like a fibre, might be very unfavorable for such purposes due to the small contact area with the substrate, which approximately corresponds just to a very thin line. However, when a fibre is thin and long enough, it may get so compliant that it comes in a close contact with the substrate over a substantial part of its length. The prerequisite for strong attachment in this case is a high toughness (or in the language of adhesion science – high cohesion) of the material, otherwise the fibre would break under applied tensile load during detachment. High adhesive forces can then be recruited by van der Waals forces and energy dissipation due to the flexibility of the fibre, as in the case of thin and strong carbon nano-tubes (Yurdumakan et al. 2005; Schaber et al. 2015a, b). In most cases there are bundles of such fibres, which altogether may behave as a highly compliant thread or pad. Although interlocking effects may play an additional role in the attachment of such fibres or fibre bundles, we here treat these in a separate chapter, since adhesive mechanisms are evident.

J.O. Wolff, S.N. Gorb, *Attachment Structures and Adhesive Secretions in Arachnids*, Biologically-Inspired Systems 7,
DOI 10.1007/978-3-319-45713-0_4

4.1 Cribellate and Aciniform Silk

The diversified silk of spiders is regarded as a key innovation of this order. It is a protein-based fibrous secretion that is produced by abdominal silk glands. The glands are differentiated by their morphology and open in seta-like tubes, the spigots, located on modified opisthosomal legs, the spinnerets. The spigots act like nozzles that shape the silk fibre. Silk is used for various purposes, like prey capture, locomotion, shelter and reproduction. These functions can only be fulfilled if there are proper means of attachment to substrates or to other threads. The attachment can be achieved by very different mechanisms, like mechanical interlocking or entangling with substrate surface structures, agglutination, or van-der-Waals, viscous/viscoelastic forces (each of these mechanisms is discussed at appropriate places in this book).

4.1.1 Cribellar Capture Thread

In some spiders, like Eresidae, Uloboridae, Deinopidae and Zoropsidae, the anterior median spinnerets are modified into flat plates, which bear numerous minute spigots. This, so called, cribellum produces bundles of nano-fibres (cribellar silk). A single fibre of cribellate silk is only 10–100 nm thick (Peters 1984). The cribellar silk fibres are arranged around a pair of thicker axial fibres of, so-called, pseudo-flagelliform silk, which stabilize the structure over a substantial length (Peters 1984) (Fig. 4.1b). Cribellar silk is used in webs for prey capture or in shelters to deter predators. It may immobilize an arthropod prey or predator by entangling with setal structures (Opell 1994). However, it can also produce significant adhesion on smooth substrates by either van-der-Waals forces (in a dry environment) or capillary forces (in a humid environment) (Hawthorn and Opell 2002, 2003). The action of electrostatic forces has repeatedly been suggested, (Peters 1986; Kronenberger and Vollrath 2015), but could not be supported by experiments (Opell 1995a).

Whereas slight ambient humidity may enhance adhesion (Hawthorn and Opell 2003), wetting highly reduces the adhesion of cribellar threads (Elettro et al. 2015), showing that it is a true dry, structure-based adhesive. On smooth surfaces, adhesion is not governed by contact area, but by thread diameter, because prior detachment the stress is concentrated at the pull-off edge (Hawthorn and Opell 2003; Opell and Schwend 2009). This is similar to the contact behavior of tapes and spatular elements (see Chap. 3), and in contrast to the viscid capture threads of the higher orb web and cob web spiders (Sahni et al. 2014). The cribellar fibrils are not smooth in many species, but exhibit tiny nodes in regular spaces (Fig. 4.1c), which arise from a specific chambered structure of the spigots (Kronenberger and Vollrath 2015). Comparison of the stickiness of smooth and noded fibrils showed that the nodes are beneficial for adhesion, presumably by generating points of very close contact with the substrate, at which intermolecular forces are effective (Hawthorn and Opell

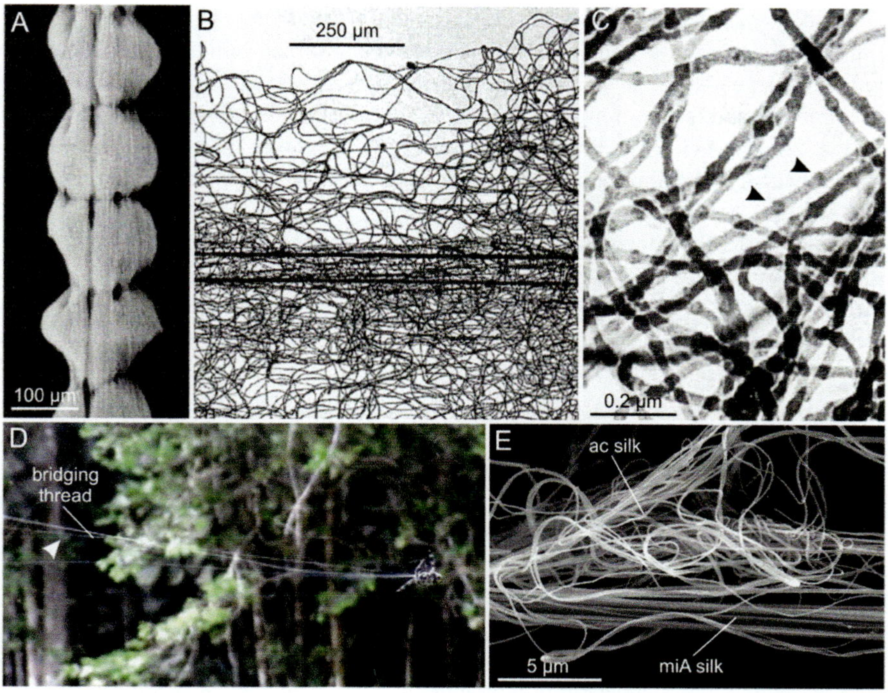

Fig. 4.1 Fibrillar pads of silk threads in spiders. (**a**) The capture threads of cribellar spiders (here: Uloboridae) are dry adhesives consisting of numerous nano-fibres brushed up with hook-like setae on the hind leg to form soft puffs (adapted from Sahni et al. 2014). (**b**) Detail of the pad of cribellar fibres around the double stranded axial thread (adapted from Peters 1986). (**c**) TEM image of the cribellar nano-fibres, showing regularly arranged globules that may help increasing the surface area and adhesion by van der Waals forces (adapted from Peters 1986). (**d**) The bridging thread used by orb web spiders to reach distant sites and initiate suspended webs similarly consist of multiple fibres (*arrowhead*); in the image a Darwin's bark spider (*Caerostris darwini*) letting a bridging thread into the air (frame of video taken in the field, adapted from Gregorič et al. 2011). (**e**) SEM image of a bridging thread from the grey cross spider (*Larinioides sclopetarius*), showing the double stranded axial thread of minor ampullate gland (miA) silk and a puff of nano-fibres of aciniform gland (ac) silk

2003). An alternative explanation of these nodes might be the contact splitting of the continuous contact line into discrete contact points, which could lead to a crack stop at every single attachment point during peeling (Kendall 2001). This effect is well known from hairy attachment pads situated on the legs of insects, spiders, and geckoes (Scherge and Gorb 2001; Varenberg et al. 2010) (see Chaps. 3 and 10). The nodes may also additionally prevent self-matting of the bundled fibres (Hawthorn and Opell 2003).

The cribellate fibres are emerging in parallel from the numerous spigots arranged in arrays. If they would stay in this arrangement they might stick together, forming a tough, but rigid bundle, which would lose its compliancy and accordingly its stickiness. Therefore the spiders brush up ('hackle') the silk with a specific structure of hook like setae in their hind legs, the calamistrum (Peters 1984; Joel et al. 2015). This results in a puffy constitution of the cribellate thread bundle (Fig. 4.1a). The puffiness depends on the morphology of cribellum and calamistrum, and the degree of combing, and has an influence on the stickiness of the entire construction (Opell 1995b, 2002). Hackling is primarily to produce this pad-like structure not to electrostatically charge the fibres as suggested by single authors (Peters 1986; Kronenberger and Vollrath 2015).

In the past it has been assumed that cribellate silk is more 'primitive' than the viscid silk of Araneidae and Theridiidae because it has been regarded as less efficient (Opell 1997). Since there are representatives of cribellate spiders that build orb webs (e.g. some Uloboridae), cribellate silk was presumed as an ancestral state in orb web spiders, because building orb webs was regarded as a highly complex behavior and its multiple evolution was stated unlikely (Coddington and Levi 1991; Opell 1997; Garb et al. 2006; Blackledge et al. 2009). However, the highly different adhesive mechanism of the capture threads in orb web spiders with cribellate (dry adhesion) and those with viscid silk (wet adhesion) have always been an issue, and the convergent evolution of orb webs has been suggested (Kullmann 1972; Elettro et al. 2015). Accordingly, more recent phylogenetic studies, using a broad set of next generation sequencing data, have revealed the multiple origin of orb webs, which means that cribellate and viscid silk are completely independent evolutionary steps towards increased web stickiness (Bond et al. 2014; Fernández et al. 2014). This is comparable to the evolution of adhesive foot pads, where two different pathways, either smooth (wet) pads (see Chap. 7) or hairy (often dry) pads (see Chap. 3) have led to the same function, namely an enhancement of foothold. In contrast to previous estimations, cribellar threads may achieve an even higher stickiness per material consumption than the viscid threads (Opell and Schwend 2009), showing that it is not functionally primitive, but a highly sophisticated product of evolution.

4.1.2 Bridging Threads and Wrapping Silk

A comparable adhesive mechanism might be present in the bridging lines of Araneidae and Linyphiidae. These are used to initiate webs or reach distant places by wind drift (Fig. 4.1d). By the use of this mechanism, even sites can be reached that are not accessible on the terrestrial way, and webs can be suspended over running waters. For instance, the giant orb web spider *Caerostris darwini* can bridge up to 25 m broad rivers by this means (Gregorič et al. 2011). The blown thread affixes at a distant structure by entanglement and mechanical interlocking, but can also stick to smooth surfaces (Wolff et al. 2014). The bridging lines are composed of minor ampullate silk and bundles of aciniform silk in the distal (first spun) part (Peters 1990; Wolff et al. 2014) (Fig. 4.1e). In large orb web spiders, like the Australian *Eriophora* spp. the bundles form a large puff at the end of the line, easily visible by eye (J.O. Wolff, pers. obs.). Aciniform silk fibres have a diameter comparable to cribellar silk and are thus very compliant. The minor ampullate silk is very stretchable (Blackledge and Hayashi 2006) and thus can partly absorb mechanical stress. Hence, its structure and properties are very similar to cribellate capture threads, proposing a similar adhesive mechanism.

Aciniform silk is also used for prey wrapping (Blackledge and Hayashi 2006) (see also Chap. 2). A tight casing is produced in order to immobilize and restrain the prey. In this case the fibre-bundle does not form a loose puff, but a casing of stressed, parallelly arranged aciniform threads. This specific type of attachment presumably works without the contribution of any glue. High friction forces may act between single aciniform fibrils due to van der Waals forces in the locally very close contact.

4.2 Particle Catching Setae

In some spiders, there are setal structures with very long and fine microtrichia, which presumably are able to generate a very intimate contact and produce adhesion due their low bending stiffness and flexibility. In *Sicarius* (Sicariidae) and *Homalonychus* (Homalonychiidae) analogously a rather similar setal morphology has evolved: the thick and stiff shaft is covered with very thin microtrichia (Fig. 4.2b), which have been hypothesized to adhere to sand particles by van der Waals forces (Roth 1984; Duncan et al. 2007). Adhering sand serves camouflage, since it blends the spider in its microhabitat (soil-crypsis) (Fig. 4.2a).

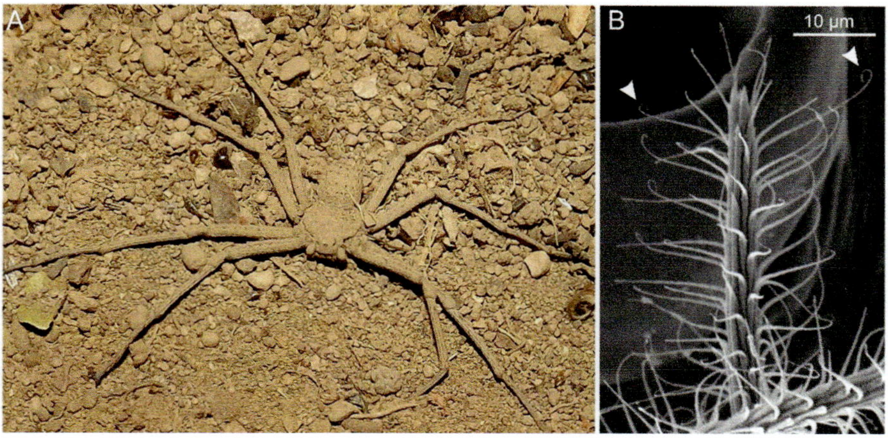

Fig. 4.2 Soil particle catching setae in Sicariidae. (**a**) The spider *Sicarius tropicus* can totally blend in its environment by a homogeneous cover with fine soil particles. (**b**) The particles adhere to setae with fine, elongated microtrichia (*arrowheads*) (here: setae of the pedipalp of *Sicarius cariri*) (Images by I. Magalhães, with kind permission)

4.3 Fibrillar Locomotory Pads Lacking Spatula

Harvestmen (Opiliones) exhibit hairy soles in their pseudo-segmented tarsi. These are composed of densely packed setae terminating in very fine filamentous tips (Fig. 4.3b, c). As discussed in Chap. 2, we assume that these enhance friction with the substrate by interlocking with minute asperities. We measured the friction angle of the short-legged *Paroligolophus agrestis* and the long-legged *Dicranopalpus ramosus* on different substrates using a tilting platform (J.O. Wolff and J. Nahrstedt, prev. unpub.). Interestingly, the hairy soles behave comparable to adhesive pads with spatulate microtrichia (see Chap. 3): the friction of the soles is even better on a smooth epoxy resin surface (84 ± 21°, N = 11 in *D. ramosus*; 94 ± 15°, N = 12 in *P. agrestis*) than compared to a micro-rough epoxy resin surface with a mean asperity diameter of 0.3 μm (58 ± 18° in *D. ramosus*; 75 ± 18° in *P. agrestis*). Friction increases slightly with increasing roughness, and gets larger than on the smooth surface, when the roughness is sufficient for the claws to interlock (mean asperity size 9 μm: 116 ± 31° in *D. ramosus*; 98 ± 25° in *P. agrestis*). This indicates that the fine setae can work as a kind of dry adhesive, although the adhesion is not sufficient to hold the animal upside down on a smooth surface. Similar frictional effects have been previously demonstrated for toe setae of chameleons having very similar shapes and dimensions to those of harvestmen (Spinner et al 2013, 2014). The harvestmen with their pseudo-segmented tarsi (especially those belonging to the sub-orders Eupnoi and Dyspnoi) can highly enhance friction by applying normal forces by bending or wrapping the leg tip around substrate structures (see Chap. 2). Analogously, chameleons have evolved gripping feet to enhance the action of friction on cylindrical substrates (small diameter branches) (Spinner et al. 2014).

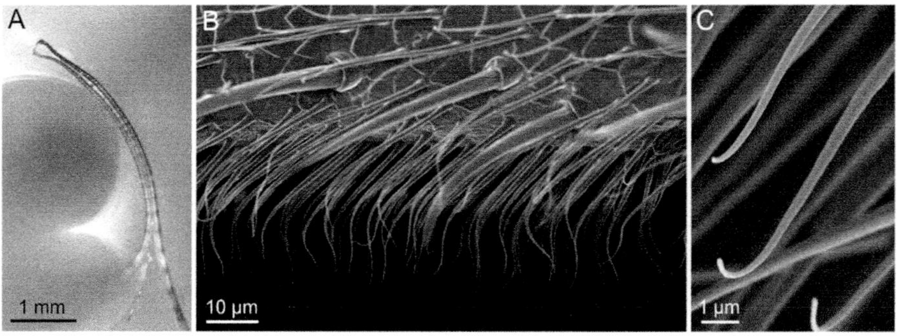

Fig. 4.3 Fibrillar tarsal pads of harvestmen. (**a**) Harvestmen can achieve considerable friction on smooth surfaces; here the distal tarsus of a tree harvestmen (*Dicranopalpus ramosus*) climbing on a glass stab. (**b**) SEM image of the tarsal sole (*lateral view*); setae with very fine, flexible tips may be responsible for friction generation on smooth and micro rough surfaces. (**c**) Detail of setal tips

References

Blackledge TA, Hayashi CY (2006) Silken toolkits: biomechanics of silk fibers spun by the orb web spider *Argiope argentata* (Fabricius 1775). J Exp Biol 209(13):2452–2461. doi:10.1242/Jeb.02275

Blackledge TA, Scharff N, Coddington JA, Szüts T, Wenzel JW, Hayashi CY, Agnarsson I (2009) Reconstructing web evolution and spider diversification in the molecular era. Proc Natl Acad Sci 106(13):5229–5234

Bond JE, Garrison NL, Hamilton CA, Godwin RL, Hedin M, Agnarsson I (2014) Phylogenomics resolves a spider backbone phylogeny and rejects a prevailing paradigm for orb web evolution. Curr Biol 24(15):1765–1771

Coddington JA, Levi HW (1991) Systematics and evolution of spiders (Araneae). Annu Rev Ecol Syst 22:565–592

Duncan RP, Autumn K, Binford GJ (2007) Convergent setal morphology in sand-covering spiders suggests a design principle for particle capture. Proc R Soc B Biol Sci 274(1629):3049–3057

Elettro H, Neukirch S, Antkowiak A, Vollrath F (2015) Adhesion of dry and wet electrostatic capture silk of uloborid spider. Sci Nat 102(7–8):1–4

Fernández R, Hormiga G, Giribet G (2014) Phylogenomic analysis of spiders reveals nonmonophyly of orb weavers. Curr Biol 24(15):1772–1777

Garb JE, DiMauro T, Vo V, Hayashi CY (2006) Silk genes support the single origin of orb webs. Science 312(5781):1762–1762

Gregorič M, Agnarsson I, Blackledge TA, Kuntner M (2011) How did the spider cross the river? Behavioral adaptations for river-bridging webs in *Caerostris darwini* (Araneae: Araneidae). Plos One 6(10):e26847

Hawthorn AC, Opell BD (2002) Evolution of adhesive mechanisms in cribellar spider prey capture thread: evidence for van der Waals and hygroscopic forces. Biol J Linn Soc 77(1):1–8

Hawthorn AC, Opell BD (2003) van der Waals and hygroscopic forces of adhesion generated by spider capture threads. J Exp Biol 206(22):3905–3911

Joel A-C, Kappel P, Adamova H, Baumgartner W, Scholz I (2015) Cribellate thread production in spiders: Complex processing of nano-fibres into a functional capture thread. Arthropod Struct Dev 44(6):568–573

Kendall K (2001) Molecular adhesion and its applications: the sticky universe. Springer Science & Business Media, Dordrecht/Boston/London, 428 pp

Kronenberger K, Vollrath F (2015) Spiders spinning electrically charged nano-fibres. Biol Letters 11(1):20140813

Kullmann EJ (1972) The convergent development of orb-webs in cribellate and ecribellate spiders. Am Zool 12(3):395–405

Opell B (1994) The ability of spider cribellar prey capture thread to hold insects with different surface features. Funct Ecol 8(2):145–150

Opell BD (1995a) Do static electric forces contribute to the stickiness of a spider's cribellar prey capture threads? J Exp Zool 273:186–189

Opell BD (1995b) Ontogenetic changes in cribellurn spigot number and cribellar prey capture thread stickiness in the spider family Uloboridae. J Morphol 224:47–56

Opell BD (1997) The material cost and stickiness of capture threads and the evolution of orb-weaving spiders. Biol J Linn Soc 62(3):443–458

Opell BD (2002) How spider anatomy and thread configuration shape the stickiness of cribellar prey capture threads. J Arachnol 30:10–19

Opell BD, Schwend HS (2009) Adhesive efficiency of spider prey capture threads. Zoology 112:16–26

Peters H (1984) The spinning apparatus of Uloboridae in relation to the structure and construction of capture threads (Arachnida, Araneida). Zoomorphology 104(2):96–104

Peters HM (1986) Fine structure and function of capture threads. In: Nentwig W (ed) Ecophysiology of spiders. Springer, New York, pp 187–202

Peters HM (1990) On the structure and glandular origin of bridging lines used by spiders for moving to distant places. Act Zool Fenn 190:309–314

Roth VD (1984) The spider family Homalonychidae (Arachnida, Araneae). Amer Mus Nov 2790:1–11

Sahni V, Dhinojwala A, Opell BD, Blackledge TA (2014) Prey capture adhesives produced by orb-weaving spiders. In: Asakura T, Miller T (eds) Biotechnology of silk. Springer, Dordrecht, pp 203–217

Schaber CF, Filippov AE, Heinlein T, Schneider JJ, Gorb SN (2015a) Modelling clustering of vertically aligned carbon nanotube arrays. Interface Focus 5:20150026

Schaber CF, Heinlein T, Keeley G, Schneider JJ, Gorb SN (2015b) Tribological properties of vertically aligned carbon nanotube arrays. Carbon 94:396–404

Scherge M, Gorb SN (2001) Biological micro- and nanotribology. Springer, Berlin/Heidelberg/New York, 309 pp

Spinner M, Westhoff G, Gorb SN (2013) Subdigital and subcaudal microornamentation in chamaeleonidae – a comparative study. J Morphol 274(6):713–723

Spinner M, Westhoff G, Gorb SN (2014) Subdigital setae of chameleon feet: Friction-enhancing microstructures for a wide range of substrate roughness. Sci Rep 4(5481):1–9

Varenberg M, Pugno NM, Gorb SN (2010) Spatulate structures in biological fibrillar adhesion. Soft Matter 6(14):3269–3272

Wolff JO, Schneider JM, Gorb SN (2014) How to pass the gap – functional morphology and biomechanics of spider bridging threads. In: Asakura T, Miller T (eds) Biotechnology of silk, vol 5. Springer, Dordrecht, pp 165–177

Yurdumakan B, Raravikar NR, Ajayan PM, Dhinojwala A (2005) Synthetic gecko foot-hairs from multiwalled carbon nanotubes. Chem Commun 30:3799–3801

Chapter 5
Mushroom-Shaped Microstructures

Abstract Mushroom-shaped microstructures exhibit a totally different contact mechanics than spatulae and are predominantly found in adhesive structures that serve long-term attachment. Among arachnids there are only few examples of such structures built by cuticle, like specialized setae in males of some mites serving attachment to the mating partner during copulation. In contrast, such structures are frequently found in external secretion products for special attachment purposes, like the anal pedicels of tortoise mites, spermatophores and spider silk thread anchors.

5.1 Introduction

Mushroom-like (or T-shaped) contact elements are composed of a circular disc with a medially attaching stalk with a thinner diameter, such that there is a radial flexible lip. Mushroom-like microstructures have been shown to generate high adhesive forces on smooth surfaces, even without a mediating fluid (Gorb et al. 2007; Carbone et al. 2011; Heepe and Gorb 2014). The generated force is comparably higher than in spatular contact elements (Spolenak et al. 2005), but detachment is comparably impeded, because the stress cannot be concentrated in a directional peeling edge. In consequence, mushroom-like structures are usually absent in locomotory attachment pads in favor of spatula-like structures (Voigt et al. 2008; Gorb et al. 2010). In contrast, such structures are often found in devices that serve a longer-term attachment, like the attachment to the mating partner, host or mother, or the (permanent) attachment of eggs, nests, threads and other secretion products (Gorb and Varenberg 2007).

Mushroom-like contact elements are usually susceptible to the pull-off angle, with a clear optimum of their action (Heepe et al. 2014; Desmond et al. 2015). Therefore there is often a constriction underneath the terminal plate, serving as a joint to reduce stress concentration at one edge of the contact. With a second joint situated at the basis of the stalk, such an effect of stress reduction can be even more enhanced.

J.O. Wolff, S.N. Gorb, *Attachment Structures and Adhesive Secretions in Arachnids*, Biologically-Inspired Systems 7,
DOI 10.1007/978-3-319-45713-0_5

5.2 Cuticular Mushroom-Shaped Microstructures

Mushroom-like microstructures built by cuticle are relatively rare in arachnids, if compared to insects. Some male mites exhibit mushroom-like adhesive setae (previously called sucker-discs) dorsally on the hind legs (Alberti and Coons 1999) (Fig. 5.1a), or, more rarely, on the front legs (Witaliński 1990). These are present in the Acaroidea (Evans 1992), e.g. in flour mites (Acaridae: *Acarus siro*), and are used to attach to females during copulation (Fain and Knülle 1981; Witaliński 1990; Alberti and Coons 1999). The setae exhibit a discoid shape, but lack a median depression and may therefore not work as a sucker. This may also count for the auxiliary suckers grouped in the adanal sucker-plate of acaridid mites (see Chap. 6).

Mites belonging to the family Nematalycidae bear single setae with mushroom-like ('cup-shaped') tips on their pedipalps, which are assumed to be used to pick up and manipulate micro-organismic prey (Bolton et al. 2015). The terminal discs are about 0.8 μm in diameter, and the authors assume a dry adhesive mechanism.

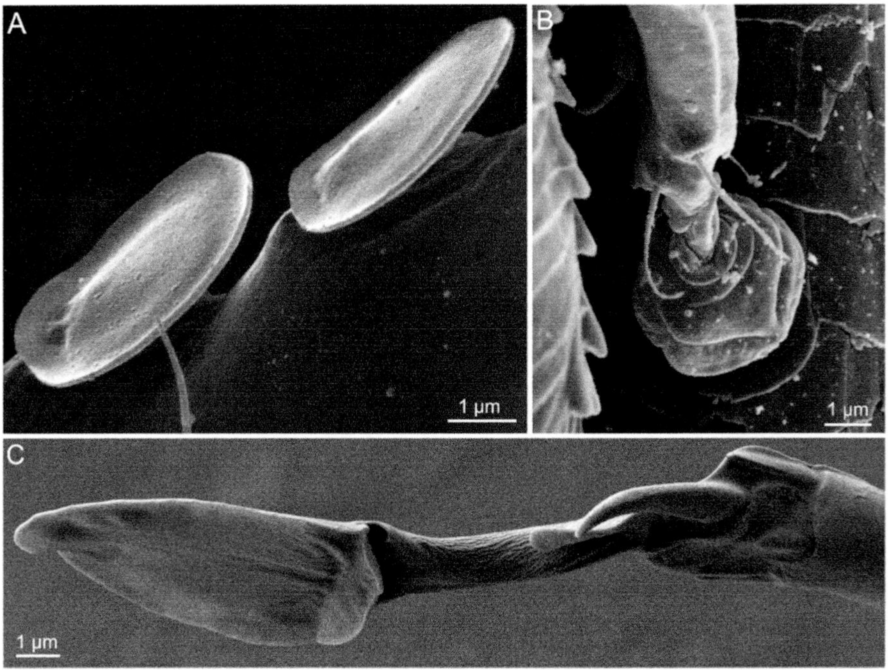

Fig. 5.1 Cuticular mushroom-like microstructures in mites. (**a**) Male acarid mites bear disc-like setae on their hind legs used to attach to the partner during copulation (after Alberti and Coons 1999, with kind permission by G. Alberti and Wiley). (**b**) In some parasitic mites the pretarsus is modified into mushroom-like adhesive pads; here a foot of *Chirodiscoides caviae* attaching to the surface of a guinea pig's hair (after Alberti and Coons 1999, with kind permission by G. Alberti and Wiley). (**c**) Tip of the hind leg of a *Histiostoma* sp. deutonymph, exhibiting a stalked arolium with a mushroom like tip

In some mites, the feet are modified into large disc like adhesive pads (Fig. 5.1b). Since the tarsus may be attached medially, such pads may to some extend act as a mushroom-like microstructure. Some mites also carry arolia or single tenent setae with mushroom like tips serving attachment to hosts (Fig. 5.1c). However, in many cases there is no perpendicular pull-off, and a main part of the adhesion mechanism may come from the high compliancy of the lip and a secretion film released in the contact (see Chap. 7).

5.3 Mushroom-Shaped Secretion Products

Mushroom-like structures are more frequently found in external attachment structures made by solidifying secretion. One well-known example is the byssus thread of mussels, which exhibits an adhesive plaque with the medially attached, tough thread (Smeathers and Vincent 1979; Bell and Gosline 1996; Desmond et al. 2015). Very similar structures can be found in arachnids, too, as the examples below illustrate.

5.3.1 Anal Stalks of Tortoise Mites (Uropodina)

The deutonymphs of Uropodina (Mesostigmata) show a unique adaptation to phoresy: They secrete a stalk of dried secretion with a mushroom-like cement disc to attach to the cuticle of phoretic hosts (usually beetles) (Bajerlein and Bloszyk 2004) (Fig. 5.2). This, so-called, pedicel emerges from anal glands and is strongly attached to the anus after drying, by another, but smaller mushroom-like patch (Bajerlein and Witaliński 2012). The stalk of the pedicel consists of parallel fibers that are surrounded by a layer of a sponge-like material, which also forms the adhesive disc (Bajerlein et al. 2013). The length of the pedicel is adjusted to the attachment site: at more exposed sites longer and thinner (hence, more flexible) pedicels are produced, in order to reduce the stress on the adhesive disc (Bajerlein et al. 2013). Furthermore, the mites preferably attach to parts of the beetles, which have surfaces that presumably facilitate a good contact and spreading of the glue (preference of smooth over hairy surfaces, and flat over curved ones) (Bajerlein et al. 2016).

The anal pedicels of Uropodina mites are remarkably analogous to the byssus threads of mussels used for an extremely robust anchorage on substrates, however, with the difference that mussels use a radial network of such threads (Smeathers and Vincent 1979; Qin and Buehler 2013). A combination of long, tough fibrils and a compliant, sponge-like cement material may be the best solution to stay attached despite of high stress impacts.

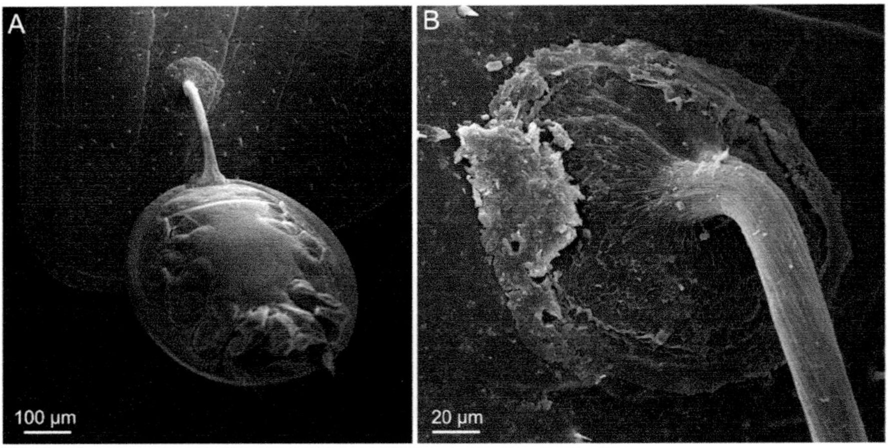

Fig. 5.2 Anal pedicel of tortoise mites (Uropodina). (**a–b**) The phoretic deutonymph of tortoise mites (here: *Uropoda orbicularis*) is a dispersive instar, which attaches to beetles by means of a stalk with an adhesive plaque of solidifying secretion (Images by Z. Adamski & D. Bajerlein, with kind permission)

5.3.2 Spermatophores

Whip-spiders, whip-scorpions, pseudoscorpions, some mites and the basal harvestmen of the sub-order Cyphophthalmi use spermatophores for sperm transfer. In all cases there is a disc-like foot conglutinated with the substrate, a flexible stalk and an apical head structure bearing the encapsulated sperm. The difference to the pedicels and byssus threads is that the stalk must be stiffer to provide an upright position. A joint-like constriction underneath the sperm carrying apical section reduces the stress acting on the adhesive disc.

In whip-spiders the spermatophores are produced in a complex glandular organ associated with the gonads and glued onto the substrate during courtship (Weygoldt 2000). We studied the spermatophores of the whip-spiders *Charon* cf. *grayi* (Charontidae) and *Phrynus longipes* (Phrynidae) (Fig. 5.3a) by means of scanning electron microscopy and found that these are composed of a loose fibrous material, covered by a thin layer of an isotropic material (Fig. 5.3b). In contrast, the adhesive plaque is disc-like and composed of an amorphous and a granular material (Fig. 5.3c). It sticks to various substrates: smooth, rough, dry and wet ones (M. Seiter, pers. comm.).

In Pseudoscorpions the spermatophore may consist of simply a sperm droplet on top of a thin stalk, which is glued onto the substrate, like in the Olpiidae, or the sperm is enclosed in a complex capsule with wing- and horn-like protuberances similar to those in amblypygids (Kew 1930; Weygoldt 1969; Legg 1973). Interestingly the spermatophores of pseudoscorpions contain fibroin (but not the same as the pseudoscorpion silk) and a lipidoid pheromon (Hunt and Legg 1971; Legg 1973). However, there are no data on the ultrastructure and adhesive mechanism of the spermatophore plaque.

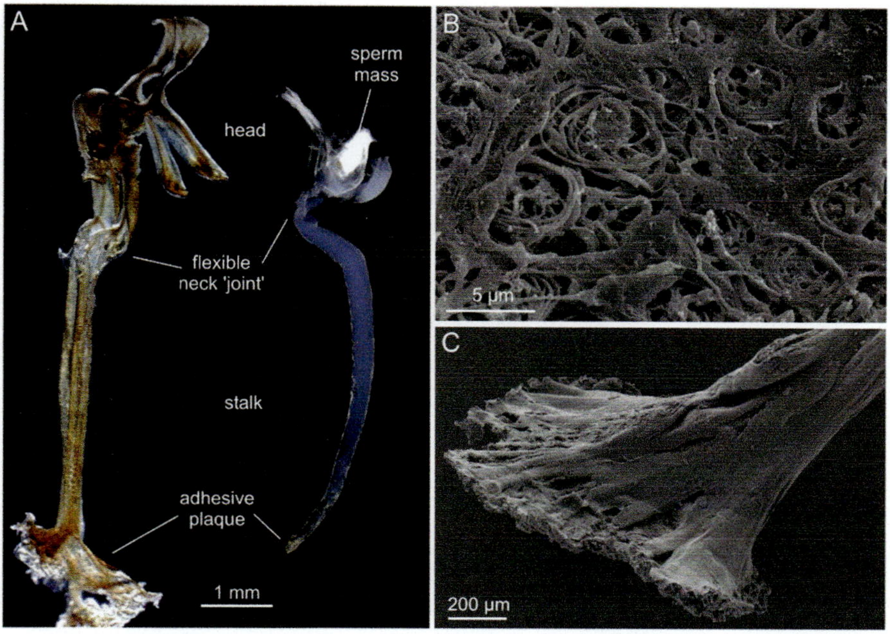

Fig. 5.3 Spermatophores of whip-spiders (Amblypygi). (**a**) Left spermatophore of *Phrynus longipes* (Phrynidae), right *Charongrayi* (Charontidae), both lateral view. (**b**) Material of the head of the spermatophore of *C. grayi*, including a fibrous and an amorphous substance. (**c**) Adhesive plaque of the spermatophore of *C. grayi*, consisting of an amorphous and a granular material

Spermatophores including an adhesive disc, a stalk and an elaborate apical part also occur in several free living acariform mite families, like Anystidae, Erythraeidae, Eriophyidae, Bdellidae (Trombidiformes), and Galumnidae (Sarcoptiformes) (Krantz and Walter 2009). Their stalk may contain a fibrous protein and is cemented onto the substrate (Kovoor 1987).

5.3.3 *Spider Silk Anchors*

Araneomorph spiders use glue-coated piriform gland silk to attach silk threads. The piriform gland silk is usually applied in disc-like structures (attachment discs) (Fig. 5.4a, b). Experimental and theoretical studies on this structure have found that load distribution, multiple elastic peeling and crack arresting synergistically work to enhance the attachment strength (Sahni et al. 2012; Pugno et al. 2013; Grawe et al. 2014; Wolff et al. 2015), which may partly reduce the demand of glue strength. This might be essential when primary gluing on plants, because strong adhesive bonds can hardly be generated on many plant surfaces due to fluid repellent properties or layers of loose wax crystals (Grawe et al. 2014). Accordingly, the attachment discs can generate high pull-off forces even on surfaces with a highly reduced free surface energy, like PTFE (Grawe et al. 2014).

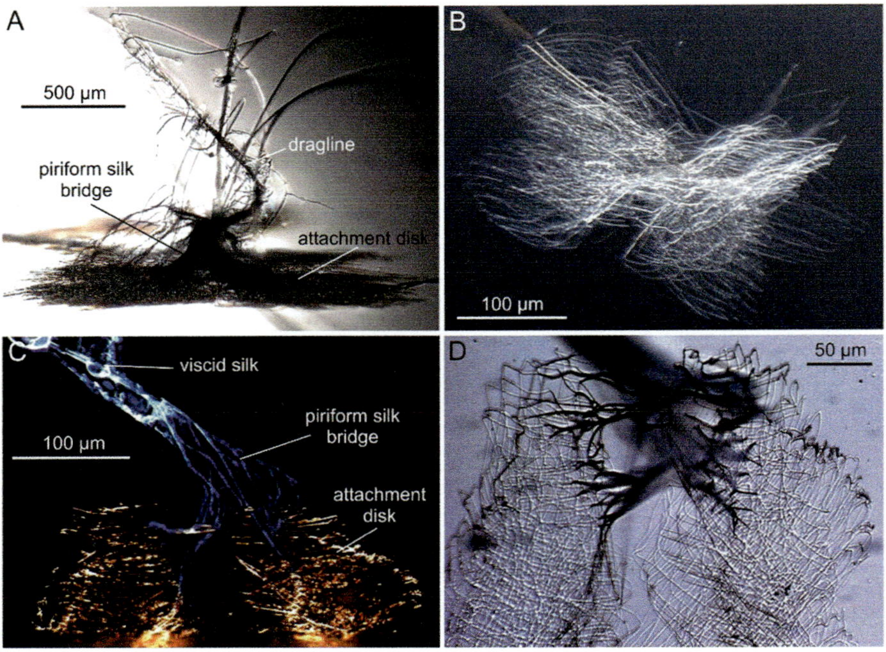

Fig. 5.4 Silk anchors of araneomorph spiders. Spiders anchor their draglines and frame threads of webs with numerous glue-coated micro-fibres from the piriform glands, applied in a mushroom like pattern. (**a**) The side view of an anchor (here of the orb weaver *Nephila senegalensis*), applied on a glass slide, shows the structure: via a flexible network of piriform gland fibres the main thread, the dragline, is medially attached to a patch of piriform gland fibres (attachment disc) agglutinated with the substrate. (**b**) Viewed from below the slide the web-like structure of the agglutinated piriform gland fibres can be seen (here an attachment disc of the hunting spider *Oxyopes heterophthalmus*). (**c**) A modified type of the attachment disc is found in so-called gumfoot-lines of cobweb spiders (here: *Parasteatoda tepidariorum*): The piriform gland silk bridge is consisting of only a few fibers presenting a defined point of breakage; silk attached to the glass appears *yellow*, whereas all silk above appears blue due to a combined illumination technique. (**d**) Same from below, note the free space in the median part

The specific architecture of the attachment disc may be controlled by the spider, in order to achieve strong or weak anchors (Sahni et al. 2012). The latter is used in the so-called gumfoot-traps of Theridiidae. These are pre-stressed threads with a viscid coating from the aggregate glands in a short section above the ground (Blackledge et al. 2005) (Fig. 5.4c). The anchorage on the ground must be so weak that it fails when a prey walks against the trap, which eventually leads to an elevation from the ground and a reduced possibility to struggle free. It was assumed that the arrangement of the fibres within the disc is used to tune the attachment strength (Sahni et al. 2012). However, our observations indicate that the bridge between the substrate cementation and the dragline is modified rather than the fibre orientation (Fig. 5.4c, d): Less piriform fibres are used to tether the dragline to the adhesive plaque, therefore it can easily break off.

References

Alberti G, Coons L (1999) Acari: mites. In: Harrison FW, Foelix RF (eds) Microscopic anatomy of invertebrates: Chelicerate Arthropoda, vol 8. Wiley-Liss, New York, pp 515–1265

Bajerlein D, Bloszyk J (2004) Phoresy of *Uropoda orbicularis* (Acari: Mesostigmata) by beetles (Coleoptera) associated with cattle dung in Poland. Eur J Entomol 101(1):185–188

Bajerlein D, Witaliński W (2012) Anatomy and fine structure of pedicellar glands in phoretic deutonymphs of uropodid mites (Acari: Mesostigmata). Arthropod Struct Dev 41(3):245–257

Bajerlein D, Witaliński W, Adamski Z (2013) Morphological diversity of pedicels in phoretic deutonymphs of *Uropodina* mites (Acari: Mesostigmata). Arthropod Struct Dev 42(3):185–196

Bajerlein D, Adamski Z, Kacalak W, Tandecka K, Wiesner M, Jurga S (2016) To attach or not to attach? The effect of carrier surface morphology and topography on attachment of phoretic deutonymphs of *Uropoda orbicularis* (Acari). Sci Nat 103:61

Bell E, Gosline J (1996) Mechanical design of mussel byssus: material yield enhances attachment strength. J Exp Biol 199(4):1005–1017

Blackledge TA, Summers AP, Hayashi CY (2005) Gumfooted lines in black widow cobwebs and the mechanical properties of spider capture silk. Zoology 108(1):41–46

Bolton SJ, Bauchan GR, Ochoa R, Klompen H (2015) A novel fluid-feeding mechanism for microbivory in the Acariformes (Arachnida: Acari). Arthr Struct Dev 44(4):313–325

Carbone G, Pierro E, Gorb SN (2011) Origin of the superior adhesive performance of mushroom-shaped microstructured surfaces. Soft Matter 7(12):5545–5552

Desmond KW, Zacchia NA, Waite JH, Valentine MT (2015) Dynamics of mussel plaque detachment. Soft Matter 11(34):6832–6839

Evans GO (1992) Principles of acarology, vol 18. CAB International, Wallingford, p 563

Fain A, Knülle W (1981) The life-cycle of *Acalvolia squamata* (Oudemans, 1909) (Acari, Astigmata, Saproglyphidae). Int J Acarol 7(1–4):139–142

Gorb SN, Varenberg M (2007) Mushroom-shaped geometry of contact elements in biological adhesive systems. J Adhes Sci Technol 21:1175–1183

Gorb S, Varenberg M, Peressadko A, Tuma J (2007) Biomimetic mushroom-shaped fibrillar adhesive microstructure. J R Soc Interface 4(13):271–275

Gorb EV, Hosoda N, Miksch C, Gorb SN (2010) Slippery pores: anti-adhesive effect of nanoporous substrates on the beetle attachment system. J R Soc Interface 7:1571–1579

Grawe I, Wolff JO, Gorb SN (2014) Composition and substrate-dependent strength of the silken attachment discs in spiders. J R Soc Interface 11(98):1742–5662

Heepe L, Gorb SN (2014) Biologically inspired mushroom-shaped adhesive microstructures. Ann Rev Mater Res 44:173–203

Heepe L, Carbone G, Pierro E, Kovalev AE, Gorb SN (2014) Adhesion tilt-tolerance in bio-inspired mushroom-shaped adhesive microstructure. Appl Phys Lett 104(1):011906

Hunt S, Legg G (1971) Characterization of the structural protein component in the spermatophore of the pseudoscorpion *Chthonius ischnocheles* (Hermann). Comp Biochem Physiol B 40(2):475–479

Kew H (1930) On the spermatophores of the pseudoscorpions *Chthonius* and *Neobisium*. Proc Zool Soc London 1930:253–256

Kovoor J (1987) Comparative structure and histochemistry of silk-producing organs in arachnids. In: Nentwig W (ed) Ecophysiology of spiders. Springer, Berlin/Heidelberg/New York, pp 160–186

Krantz GW, Walter DE (2009) A manual of acarology, 3rd edn. Texas Tech University Press, Lubbock, 816 pp

Legg G (1973) Spermatophore formation in the pseudoscorpion *Chthonius ischnocheles* (Chthoniidae). J Zool 170(3):367–394

Pugno NM, Cranford SW, Buehler MJ (2013) Synergetic material and structure optimization yields robust spider web anchorages. Small 9(16):2747–2756

Qin Z, Buehler MJ (2013) Impact tolerance in mussel thread networks by heterogeneous material distribution. Nat Comm 4:2187

Sahni V, Harris J, Blackledge TA, Dhinojwala A (2012) Cobweb-weaving spiders produce different attachment discs for locomotion and prey capture. Nat Comm 3:1106

Smeathers J, Vincent J (1979) Mechanical properties of mussel byssus threads. J Molluscan Studies 45(2):219–230

Spolenak R, Gorb S, Gao H, Arzt E (2005) Effects of contact shape on the scaling of biological attachments. Proc R Soc Lond A 2054:305–319

Voigt D, Schuppert JM, Dattinger S, Gorb SN (2008) Sexual dimorphism in the attachment ability of the Colorado potato beetle *Leptinotarsa decemlineata* (Coleoptera: Chrysomelidae) to rough substrates. J Insect Physiol 54:765–776

Weygoldt P (1969) Biology of pseudoscorpions. Harvard University Press, Cambridge/Massachusetts

Weygoldt P (2000) Whip spiders (Chelicerata: Amblypygi): their biology, morphology and systematics. Apollo Books, Stenstrup, 160pp

Witaliński W (1990) Adanal suckers in acarid mites (Acari, Acaridida): Structure and function. Int J Acarol 16(4):205–212

Wolff JO, Grawe I, Wirth M, Karstedt A, Gorb SN (2015) Spider's super-glue: thread anchors are composite adhesives with synergistic hierarchical organization. Soft Matter 11(12):2394–2403

Chapter 6
Suction Cups

Abstract Suction cups can produce reversible attachment by creating a pressure gradient. Such devices are usually restricted to aquatic animals. Among arachnids suckers are present in the mite orders Sarcoptiformes and Trombidiformes. These are remarkable, because they occur in primarily terrestrial animals, and with diameters of often below 10 μm they are smaller than the sucker efficacy threshold determined in previous theoretical models. So-called adanal suckers are situated at the ventral posterior side and used to attach to phoretic hosts, to reach new habitats, or in males to attach to the mating partners during long-term copulations, in order to secure paternity. Many acaridid mites possess different types of suction cups grouped together on a plate, that itself works like a suction cup and is filled with a viscous, non-volatile (presumably oily) secretion. Sucker-like structures are also found in the mouth parts of some mites, used to attach to their hosts while feeding.

6.1 Introduction

Suction cups are structures that produce a pressure difference between a cavity, which is filled with either a gas (e.g. air) or a liquid (e.g. water), and the surrounding. The force, a suction cup can exert, is basically related to the pressure gradient, i.e. the difference between the pressure outside the cup and that inside the cup, times the area of the surface covered by the cup. Thus, the higher the pressure difference, the higher is the attraction between the sucker and the substrate.

It is useful to differentiate between suction cups that operate with a gas or a liquid. Suppose a suction cup is used in air under dry, ambient conditions. Then, the maximum adhesive strength, maximum attractive force divided by the area of the surface covered by the cup, is limited to 1 atm (~100,000 Pa). By increasing the inner volume of the cup by e.g. retraction of the median part, the inner pressure is continuously lowered. In different animals, the retraction can be performed by muscles and/or the elastic energy of the initially deformed cup (Kier and Smith 1990;

Electronic supplementary material The online version of this chapter (doi:10.1007/978-3-319-45713-0_6) contains supplementary material, which is available to authorized users Videos can also be accessed at http://www.springerimages.com/videos/978-3-319-45712-3.

J.O. Wolff, S.N. Gorb, *Attachment Structures and Adhesive Secretions in Arachnids*, Biologically-Inspired Systems 7,
DOI 10.1007/978-3-319-45713-0_6

Frutiger 2002; Gorb 2008). However, in reality due to surface irregularities and contaminations there is always a leakage current of fluid (gas or liquid) from outside to the inner side of the cup compensating for the reduced inner pressure. This drastically affects the efficiency of the system. In order to produce a stable pressure gradient, the inner cavity must be sealed as good as possible, which means that either the surrounding lip-like structure should be highly compliant (Tramacere et al. 2014a) and/or a viscous fluid or mucous acts as a sealing agent (Tramacere et al. 2014b). Suction cups in a liquid medium, in addition to a better sealing efficiency, have two more peculiarities: (i) only a very small volume change is necessary to produce a rather large pressure reduction due to the small compressibility of liquids (Smith 1991); and (ii) since liquids can withstand even tensile load, the maximum adhesive strength can exceed the value of one atmosphere and only depends on the cohesive strength of the liquid itself (Smith 1991; Caupin and Herbert 2006). Published examples of suction devices exceeding the pressure gradient from 1 atm are that of octopus suckers (Smith 1991) and of microscopic mushroom-shaped adhesive structures mimicking the hairy attachment devices found in some insects (Heepe et al. 2014).

In contrast to adhesive pads the pull-off force of a suction cup scales with the area and thus with the radius squared $\sim r^2$ (Spolenak et al. 2005). The pull-off force of adhesive pad structures (here assumed to rely on intermolecular van der Waals interaction) scales slower $\sim r^{<2}$ with the radius of contact (Spolenak et al. 2005). One can thus define two regimes, a van der Waals-dominated regime and a suction-dominated regime, depending on the size of the contact. Below that critical size this scaling suggests that it is more effective to rely on intermolecular van der Waals forces rather than on suction. Above that value it is the other way round. The size of contact at which the transition from one regime to another occurs can be estimated to be of order a few micrometers. However, this scaling rule does not account for surface irregularities, contaminations, or potential fluids sealing the cavity. Thus the exact value of the critical size separating those two regimes may strongly vary and depends on many external factors.

Among arachnids, suction cups are only known from mites (Sarcoptiformes and Trombidiformes). These suckers are minute, belonging to the smallest suction cups known. Furthermore, most of the mites bearing these suckers are terrestrial. This renders these structures very interesting for micro-mechanic studies due to additional factors enhancing their performance in a non-aquatic environment.

6.2 Adanal Suckers in Mites

Suckers independently evolved in the anal region of astigmatid mites (Sarcoptiformes: Acaridida) and gall mites (Trombidiformes: Eriophyidae) (Popp 1967; Baker et al. 1987; Witaliński 1990; Alberti and Coons 1999).

In the Acaridida the suckers occur in phoretic nymphs (so-called heteromorphic deutonymphs or *hypopi*) (Alberti and Coons 1999; Chmielewski 2009) and in males

(Popp 1967; Witaliński 1990). In the first case they serve the attachment to arthropod phoretic hosts in order to disperse, in the latter case they fulfill the function of partner fixation, which plays a role for mate guarding (Witaliński 1990; Witaliński et al. 1992). Both the adanal suckers of male acarids and the main suckers of hypopi have a concave shape and medially attached retractor muscles, which indicates an active sucker function (Baker et al. 1987; Witaliński 1990). In the slightly depressed center, they bear a tiny nipple-like structure (see Figs. 6.1f, h), which is connected to dendrites of a sensory cell and assumed being a mechanoreceptor (Witaliński 1990).

The suckers of male acarids are assumed to be modified setae, and are located on both sides of the anal slit, i.e. in flour mites (*Acarus siro*) (Griffiths et al. 1990; Witaliński 1990). The suckers are usually slightly depressed in a button-like fashion, but may also be tube-like with a piston-like moveable internal structure, which can highly change the inner volume (Witaliński 1990; Witaliński et al. 1992). The structure of the suction cups is described as rigid and stable in shape, formed by dense exocuticle. In contrast, the margins are very soft and compliant and made of less dense procuticle (Witaliński 1990). When pressed onto the substrate (the female cuticle), the soft margins likely seal the inner cavity, which is then increased by muscular retraction of the rigid central part (Witaliński 1990). The above author assumes that by this means vacuum is produced, which, however, is unlikely as discussed above. Presumably, they are filled by a fluid secretion, as we found in the hypopal suckers (see below). Interestingly, the target surfaces on the female are often conspicuously smooth, in contrast to the often microsculptured cuticle on the rest of the body (Witaliński et al. 1992). This is obviously to ensure a proper function of the suction cups. In feather mites (Proctophyllodidae) the female even possesses tubercles that fit into the modified tube-like suckers of the males (Witaliński et al. 1992). These tubercles can be slightly inflated by an increase of the internal hemolymph pressure, which presumably enhances the sealing of the sucker and hence increases the efficacy of the attachment (Witaliński et al. 1992).

In the hypopi, suction cups of different size and structure are grouped on an adanal (means: next to the anal opening) sucker plate (Baker et al. 1987). This plate is sclerotized and slightly convex, like a sucker itself. Some suckers, like the auxiliary suckers surrounding the movable main suckers (Fig. 6.1c) are very flat and not suitable to produce a cavity. Their function might be more related to passive suction devices similar to the mushroom-shaped contact elements (see Chap. 5). The exact mechanism of these structures demands closer investigations.

Based on the potentially restricted functionality of suckers under dry, ambient conditions (see above) one may hypothesize that some fluid secretions may be involved. Glands have been found near the sucker plate (Popp 1967), but if and how a secretion emerges into the suckers was unknown.

We did some investigations on hypopi of *Histiostoma* sp. (Histiostomatidae), which are phoretic on earwigs (Dermaptera) (Chmielewski 2009). By means of reflection interference contrast microscopy (RICM) we found that the mites attach exclusively with their sucker plates, which are filled with a viscous, non-volatile fluid (Figs. 6.1b, d; Videos 6.1 and 6.2). Mites could hardly detach on their own

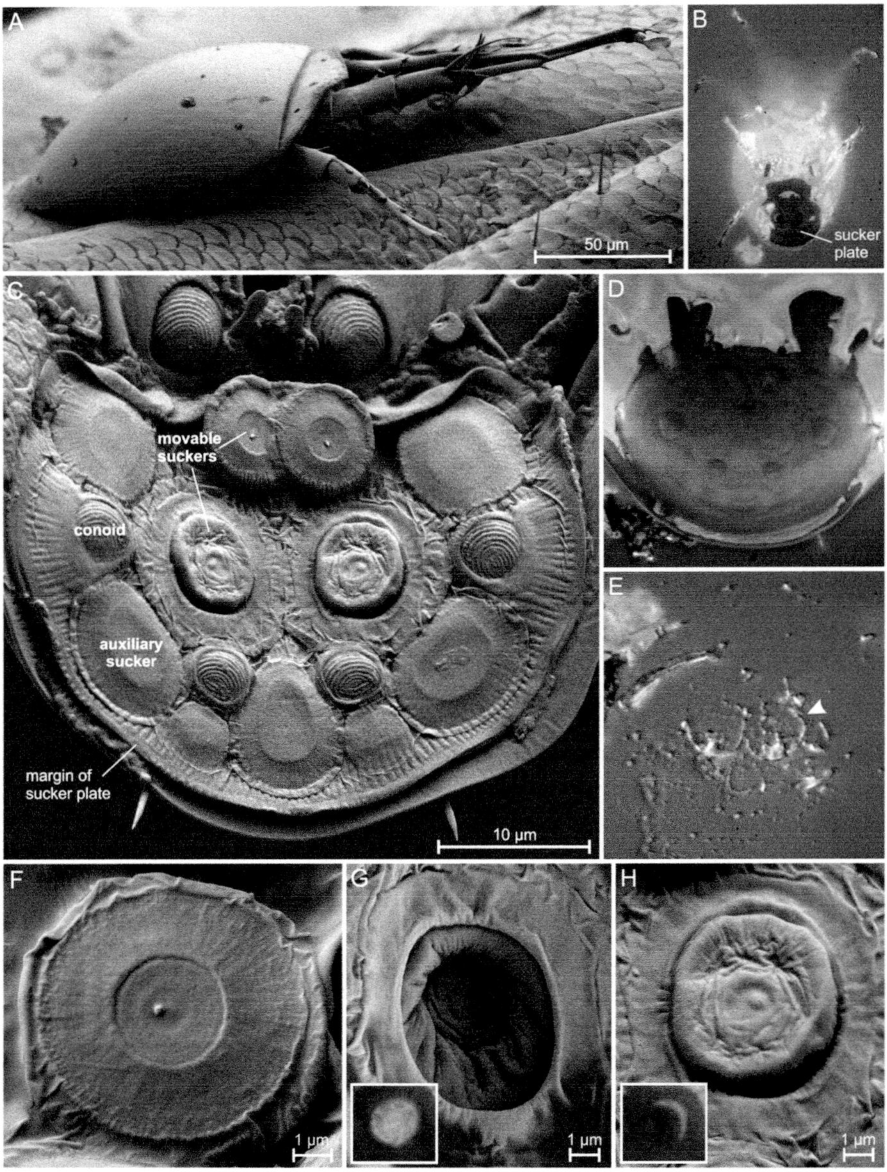

Fig. 6.1 Adanal sucker plate of phoretic mite nymphs (hypopi). All images show hypopi of *Histiostoma* sp. (Sarcoptiformes: Histiostomatidae), obtained by cryo-scanning electron microscopy, and (**b**), (**d**) and insets in (**g**) and (**h**) by reflection interference contrast microscopy (RICM). (**a**) Hypopus attached to the dorsal cuticle of its phoretic host, the common earwig *Forficula auricularia*. (**b**, **d**) RICM image of the same attached to a cover slip, viewed from below. This method reveals that the sucker plate is filled with a viscous secretion. (**c**) Sucker plate at the posterior ventral part of the hypopi. Suckers and the ribbed conoids are assumed to be modified setae. Secretions seem to come from the conoids. Three types of suckers are present, with the median

from the glass slide, although they use action of their legs for this purpose. Only a single mite out of five mites tested detached itself, and in this case remains of the secretion were left behind, resembling the shapes of the median suckers (Fig. 6.1e; Video 6.1). Histiostomatid mites have been shown to carry a coat of an oily secretion, which is typical for most sarcoptiform mites (Koller et al. 2012). The secretion of the sucker plate is presumably of a similar composition to the cuticle grease, since similar fluid remains were deposited, when other body parts touched the glass slide. We could observe that both the median main suckers and the auxiliary suckers are capable to seal the cup cavity with air enclosure in some cases. The median suckers can produce the largest cavity, and active pro- and retraction of the sucker was observed.

In gall mites (Eriophyidae) the sucker is used to attach to plant surfaces, to phoretic hosts (usually insects), to each other, in order to build chains that are drifted by wind currents, but also for looper-like locomotion and even leaping (Baker et al. 1987). The sucker is simply a softening of the cuticle around the anus by a lack of sclerotization and the reduction of epi- and exocuticle (Baker et al. 1987). It is inflated by the internal hemolymph pressure. Due to the high compliancy of the membranous cuticle, the inner cavity of the anus is sealed, when pressed against the substrate. The anus then can be retracted by attached muscles, which produces a drop of pressure (Baker et al. 1987). During molting, attachment is enhanced by the secretion of a viscous fluid emerging from the anal opening (Baker et al. 1987). It remains unclear, to which extent a suction effect plays a role in the adhesion of this structure. The soft margin may already function as a smooth adhesive pad, especially if fluids are involved.

6.3 Sucker-Like Mouthparts

Another type of suckers occurs in some water mites like *Limnochares aquatica* (Limnocharidae), whose larvae are ectoparasites of water insects: The gnathosomal tip has a round suction cup like depression with the mouthparts in its middle (Fig. 6.2a) (Alberti and Coons 1999). The margin consists of non-slerotized cuticle

Fig. 6.1 (continued) ones actively movable by muscles. (**e**) Secretion remains after detachment of the hypopi. Note the imprints of the pair of median main suckers (*arrowhead*) and the posterior auxiliary sucker. See Video 6.1 for motions of the sucker plate during detachment. (**f**) Elevated movable sucker in the anterior part of the plate. Note the very thin and flexible cuticle around the sucker, whereas the sucker itself appears to be stiff and stable in shape. (**g–h**) Median main sucker, (**g**) retracted, (**h**) protracted. Insets show images of a RICM-HSV sequence, in which the mite pro- and retracted the sucker (see Video 6.2) (Images by L. Heepe and JO. Wolff)

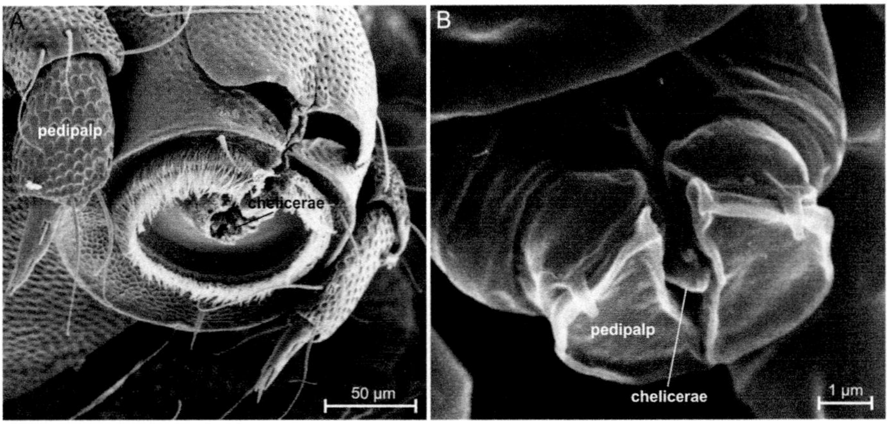

Fig. 6.2 Sucker-like mouth parts in mites. (**a**) Sucker formed by the gnathosomal tip of the water mite *Limnochares aquatica* (Trombidiformes: Limnocharidae). (**b**) Mouth sucker of the gall mite *Phytoptus avellanae* (Trombidiformes: Sierraphytoptidae). Note that in this case the sucker is formed by the pedipalps (Images after Alberti and Coons 1999, with kind permission by G. Alberti and Wiley)

with fine microtrichia, between which a mucous fluid might be secreted. Due to the limnic environment, a significant sucker function to attach to the host's cuticle is probable.

Also the mouthparts of gall mites are modified into a sucker-like structure. In this case the sucker is not formed by the gnathosomal cuticle but rather by the flattened tips of the pedipalps that enclose the needle-like chelicerae (Fig. 6.2b). The pedipalps are used to attach to the plant surface, while piercing it and feeding on the plant sap.

Acknowledgements Dr. Lars Heepe (Functional Morphology and Biomechanics, Zoological institute, Kiel University, Germany) helped in obtaining the original data and writing this chapter.

References

Alberti G, Coons L (1999) Acari: mites. In: Harrison FW, Foelix RF (eds) Microscopic anatomy of invertebrates: Chelicerate Arthropoda, vol 8. Wiley-Liss, New York, pp 515–1265

Baker G, Chandrapatya A, Nesbitt H (1987) Morphology of several types of cuticular suckers on mites (Arachnida, Acarina). Spixiana 10:131–137

Caupin F, Herbert E (2006) Cavitation in water: a review. Comptes Rendus Physique 7(9):1000–1017

Chmielewski W (2009) Phoretic mites (Acarina) on earwigs, *Forficula auricularia* L. (Insecta, Dermaptera), found in apiaries. J Apicultural Sci 1(53):75–80

Frutiger A (2002) The function of the suckers of larval net-winged midges (Diptera: Blephariceridae). Freshwater Biol 47(2):293–302

Gorb SN (2008) Biological attachment devices: exploring nature's diversity for biomimetics. Phil Trans R Soc A 366(1870):1557–1574

Griffiths D, Atyeo W, Norton R, Lynch C (1990) The idiosomal chaetotaxy of astigmatid mites. J Zool 220(1):1–32

Heepe L, Kovalev AE, Gorb SN (2014) Direct observation of microcavitation in underwater adhesion of mushroom-shaped adhesive microstructure. Beilstein J Nanotech 5(1):903–909

Kier WM, Smith AM (1990) The morphology and mechanics of octopus suckers. Biol Bull 178(2):126–136

Koller LM, Wirth S, Raspotnig G (2012) Geranial-rich oil gland secretions: a common phenomenon in the Histiostomatidae (Acari, Astigmata)? Int J Acarol 38(5):420–426

Popp E (1967) Die Begattung bei den Vogelmilben *Pterodectes* Robin (Analgesoidea, Acari). Z Morph Ökol Tiere 59(1):1–32

Smith AM (1991) Negative pressure generated by octopus suckers: a study of the tensile strength of water in nature. J Exp Biol 157(1):257–271

Spolenak R, Gorb S, Gao H, Arzt E (2005) Effects of contact shape on the scaling of biological attachments. Proc R Soc A 461(2054):305–319

Tramacere F, Kovalev A, Kleinteich T, Gorb SN, Mazzolai B (2014a) Structure and mechanical properties of *Octopus vulgaris* suckers. J R Soc Interface 11(91):20130816

Tramacere F, Appel E, Mazzolai B, Gorb SN (2014b) Hairy suckers: the surface microstructure and its possible functional significance in the *Octopus vulgaris* sucker. Beilstein J Nanotech 5(1):561–565

Witaliński W (1990) Adanal suckers in acarid mites (Acari, Acaridida): structure and function. Int J Acarol 16(4):205–212

Witaliński W, Dabert J, Walzl MG (1992) Morphological adaptation for precopulatory guarding in astigmatic mites (Acari: Acaridida). Int J Acarol 18(1):49–54

Chapter 7
Soft Adhesive Pads

Abstract Compliant pads based on a specific foam-like or fibrillar internal structure of material are a widespread type of adhesive structures made of cuticle. Such pads usually have a smooth surface at a macroscale and are supplemented by lipid-bearing secretions. Besides, in whip-spiders there is the special case of a soft pad with spatula-like microstructures directly connected with the surface. Soft pads usually occur on the pretarsus as so-called arolia, which serve dynamic attachment during locomotion and static attachment during resting. Arachnids have arolia of different shape, which influences their contact mechanics. There are cushion- or bubble-like pads, disc- or mushroom-like ones, and arolia with a stiff trunk and a broad soft lip (transverse-lip arolium). The latter type has evolved three times independently among arachnids, with a strikingly similar structure. The specific internal structure of soft adhesive pads allows for different detachment mechanisms, like invagination, folding and lateral peeling. In addition to the literature review, this chapter includes the first ultrastructural and functional original data on the arolia of pseudoscorpions, harvestman nymphs, and some mites.

7.1 Introduction

Many, especially small, arthropods have pad-like adhesive organs, which are usually located at the tip of appendages and used for attachment to the substrate during locomotion, prey capture or copulation. The main functional principle of adhesive pads is their high compliancy (Gorb 2001). Such pads are preferably soft, in order to adapt to the topography of the substrate and create a high contact area. In so-called smooth pads, the outer-most layer is very thin and thus deformable very locally under compression (Jiao et al. 2000), and at the same time tough to resist high shear forces (Bennemann et al. 2014). A mechanical stability at simultaneous maintenance of compliancy is usually achieved by a foam-like or fibrillar structure of the underlying material layer (Gorb et al. 2002; Perez Goodwyn et al. 2006;

Electronic supplementary material The online version of this chapter (doi:10.1007/978-3-319-45713-0_7) contains supplementary material, which is available to authorized users Videos can also be accessed at http://www.springerimages.com/videos/978-3-319-45712-3.

J.O. Wolff, S.N. Gorb, *Attachment Structures and Adhesive Secretions in Arachnids*, Biologically-Inspired Systems 7,
DOI 10.1007/978-3-319-45713-0_7

Schargott et al. 2006). Such arrays of fibrils are usually not perpendicular to the outer membrane, which causes an anisotropic friction of the pad (usually high friction when dragged towards the body and low friction when pushed away from the body) (Gorb 2000, 2007; Bullock et al. 2008). This is a frequent means of controlling attachment by leg positioning. Another way of adapting the attachment strength is the control of the pad contact area by pretarsal muscles (Federle and Endlein 2004). Substantial for a good adhesion of smooth pads is the secretion of a thin, continuous fluid film into the contact, which produces high capillary and viscous forces (Jiao et al. 2000; Drechsler and Federle 2006; Barnes 2007).

The smooth adhesive pads of most arachnids obtain their compliancy by a fibrillation of the underlying cuticle, just like in many insects. In the analogue pads, the fibrils similarly do not run perpendicular to the epicuticle, but are tilted distally at some angle. This may be important for anisotropy of friction generation as mentioned above, and may be simply a result of the cuticular formation, as the fibrils originate from horizontally stacked cuticular layers.

7.2 Cushion-Like Arolium

The cushion-like arolium is more or less globular, when fully extended, appearing like a bubble at the foot tip. This type is usually inflated by internal hemolyph pressure and retracted by the elasticity of surrounding structures or/and an attached flexor muscle. Accordingly, the pad surface often exhibits an accordion-like structure with regular folds to allow a high increase of its volume during inflation.

7.2.1 Prenymphal Arolium in Scorpions

Scorpions are viviparous and show maternal care. The born instar, the prenymph, is not fully developed yet and climbs onto the back of the mother, where it attaches (Millot and Vachon 1949) (Fig. 7.1a). For this purpose, the distal leg segment, the pretarsus, exhibits a large arolium at this developmental stage (Millot and Vachon 1949). This arolium is simply the non-sclerotized median part of the pretarsus (Fig. 7.1b). The ventral cuticle of the arolium is highly compliant due to an internal disaggregation into parallel fibers (Wolff et al. 2015a) (Fig. 7.1d). Its adhesion may be enhanced by the capillary forces of a secreted fluid film, which appears to be hydrophobic (Wolff et al. 2015a) (Video 7.7). Detachment is achieved by the invagination of the arolium using the attaching pretarsus retractor muscle (Wolff et al. 2015a) (Videos 7.2, 7.3, 7.4 and 7.5.).

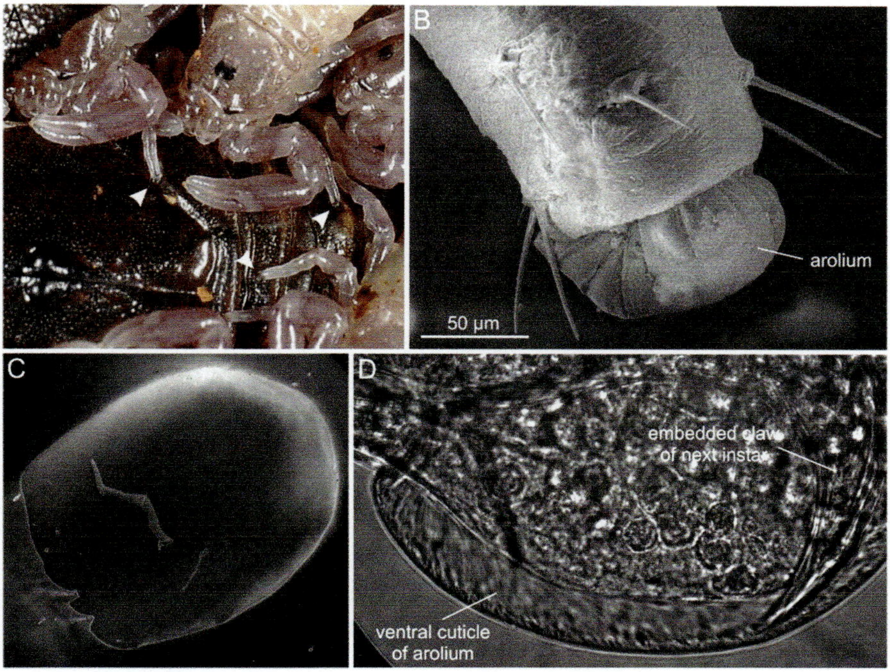

Fig. 7.1 Arolium of scorpion prenymphs. (**a**) The prenymphs of scorpions (here: *Liocheles australasiae*) attach to their mother's back with simple inflatable arolia (*arrowheads*). (**b**) The arolium is basically a cushion-like median extension of the pretarsus that can be retracted into the tarsus by a dorsally attached muscle; this image shows the lateral view of the leg tip, as visualized in the Cryo-SEM on a shock frozen prenymph. Also see Videos 7.2, 7.3, 7.4 and 7.5. (**c**) The ventral cuticle of the arolium is very soft and coated by a lipid-bearing fluid; its contact area with a smooth glass surface is oval, as seen here by means of reflection interference contrast microscopy through a cover slip. Also see Video 7.6. (**d**) A differential interference contrast micrograph of a squash preparation shows that the ventral cuticle is bloated and contains a fibrillary structure as typical for smooth adhesive pads; inside the arolium, hemolymph cells and the developing claws, which become functional after the first molting, are visible

7.2.2 Nymphal Arolium in Harvestmen (Opiliones: Laniatores)

An arolium is present in the two posterior leg pairs of nymphs in the Grassatores, which comprise the majority of Laniatores (Gnaspini 2007). The arolium is present in all nymphal instars, but absent in the larva, the subadult and adult (Gnaspini et al. 2004; Townsend et al. 2009). This organ is used to attach to overhanging surfaces during the molting process (Juberthie 1972). Molting in a hanging position presumably helps these animals to pull their long legs out of the exuvia. Furthermore, the risk of predation might be reduced by this means. With the help of the arolium, the nymph can even attach to smooth surfaces like glass panes (Juberthie 1972). Roewer (1935) compared the arolia with the pedipalpal organ of Solifugae (see below) and assumed a similar adhesive function.

We studied the ultrastructure of the nymphal pretarsus in *Paranonychus* sp. (Triaenonychidae) (Fig. 2.1o), *Holoscotolemon querilhaci* (Fig. 7.2a-c), *H. jaqueti* and *H. naturae* (Cladonychiidae), *Gnomulus* sp. (Sandokanidae) (Fig. 2.1l), *Scotolemon doriae* (Fig. 7.2f, i, j), *S. lucasi* (Fig. 7.2e, g, l), *Sitalcina californica* (Fig. 7.2k) (Phalangodidae), *Dibunus similis* (Epedanidae) (Fig. 7.2m, n), *Cynortellana quadrimaculata* (Cosmetidae) (Fig. 7.2o), *Hinzuanius flaviventris* (Biantidae) (Fig. 7.2r) and an unidentified species of Assamidae (Fig. 7.2p, q).

The front legs (first and second leg pair) of all instars in all species bear a slender hook-like single claw (Fig. 7.2a), however, the morphology of hind legs (third and fourth leg pair) differs enormously between different lineages and instars. In the rather basal *Paranonychus* sp. the nymph does not bear an arolium, but the single claws of the hind legs exhibit a spoon-like shape with lateral denticles (Fig. 2.1o). In *Holoscotolemon* spp. (Cladonychiidae), the hind leg pretarsi are highly modified in the nymphs. The median apical part is modified into an arolium like cushion, which is rather slender and distally tapered (Fig. 7.2b, c). It seems to be apically attached to the large empodium, and additionally bears numerous claw-like sclerites (auxiliary claws) on the lateral sides. We could not observe an adhesive function in nymphs of *H. naturae* – the contact area is very small on smooth glass and the nymphs were not capable to attach to inverted smooth surfaces. Hence, we assume that the pad has to get somehow activated by inflation, which is not performed during usual locomotion. Unfortunately, we could not observe a molting process to verify this hypothesis.

In the majority of derived Laniatores (clade Grassatores), the nymphs bear a comparably large arolium. Basically, the structure is rather similar between species. It is a membranous dorso-median outgrowth of the elongated pretarsus, laterally flanked by the claws. A median claw (empodium) may also be present at the basis, as seen in Assamidae (Fig. 7.2p) and Cosmetidae. The surface of the arolium is

Fig. 7.2 (continued) ventral view; (**c**) leg 4, lateral-apical view); (**d**) Hind legs of sub-adults and adults exhibit a totally different pretarsal morphology, with an overall sclerotization and two main claws only, slightly tilted towards each other (here: leg 4 of adult). (**e–l**) The arolium of Phalangodidae is broad and dorsally highly inflatable due to an accordion-like folded surface, whereas the ventral cuticle appears rather stiff; the empodium and auxiliary claws are lacking, (**e**) *Scotolemon lucasi*, leg 4 of nymph, ventral view, (**f**) *Scotolemon doriae*, leg 4 of nymph, detail of non-inflated arolium, distal edge, (**g**) *S. lucasi*, leg 4 of nymph, lateral-apical view, (**h**) *S. lucasi*, leg 4 of adult, ventral view, (**i**) *S. doriae*, leg 4 of nymph in normal state, lateral view, (**j**) same, inflated, (**k**) *Sitalcina californica*, leg 4 of nymph with inflated arolium, apical view, (**l**) Detail of folds of the inflated arolium of *S. lucasi*, apical view. (**m–r**) The arolium of Grassatores is globular and exhibits dorsal folds, a small empodium is often present; (**m**, **n**); (**m**) *Dibunus similis* (Epedanidae), leg 3 of nymph, lateral-dorsal view, note that in this species auxiliary claws are present; (**n**) A fracture of the arolium (same species) reveals the fibrillation of the ventral cuticle and the hollow inside; (**o**) *Cynortellana quadrimaculata* (Cosmetidae), leg 4 of nymph, lateral view; (**p**, **q**) Nymph of Assamiidae, unidentified species, leg 4 with inflated arolium, lateral view, (**p**) Light microscope image with the empodium visible, (**q**) Detail of unfolded dorsal cuticle of arolium; (**r**) Nymph of *Hinzuanius flaviventris* (Biantidae), leg 3 with folded arolium, lateral view

Fig. 7.2 Nymphal arolium in Laniatores harvestmen. In many harvestmen of the sub-order Laniatores the nymphs bear inflatable arolia in their hind legs, presumably for attaching to over-hanging surfaces to assist molting in a hanging position. (**a–d**) *Holoscotolemon querilhaci* (Cladonychiidae); (**a**) Legs 1 and 2 bear single hook-like claws in both nymphs and adults in Laniatores (here: leg 1 of nymph, ventral view); (**b**, **c**) Pretarsi of legs 3 and 4 of nymphs exhibit the main claws, a large empodium, auxiliary claws and a slender dorso-median arolium ((**b**) leg 3,

longitudinally wrinkled, which permits a large expansion of the pad. The folding morphology differs between some lineages (Fig. 7.2l, q). The ventral cuticle internally exhibits branched fibrils as we observed in a fractured arolium of *D. similis* (Fig. 7.2n). In *H. flaviventris,* the arolium is rather small (Fig. 7.2r). In the Phalangodidae, which are the sister group to all other Grassatores (Sharma and Giribet 2011), the arolium does not expand frontally, as in the other Grassatores, but dorsally (Figs. 7.2e, g, i–k). The ventral cuticle appears to be rather stiff; it is smooth in *S. californica*, corrugated in *S. doriae* and nubby in *S. lucasi.* In contrast, the distal and apical parts are wrinkled and highly inflatable (in some of the studied specimens, the tip was expanded, see Figs. 7.2j, k). In the other Grassatores, the membranous area extends more to the ventral part, such that it may make a contact with the substrate during normal locomotion. However, the use of arolia during locomotion remains unknown, as it was not possible to obtain living nymphs of Grassatores for this study. In the Sandokanidae, whose phylogenetic position is unclear, the arolium has totally been lost (Gnaspini 2007) (Fig. 2.11).

7.2.3 *Camel Spiders*

The pretarsus of solifuges exhibits a pro-ventral pad-like arolium (Kästner 1933), which may be subdivided in several lobes (Klann 2009). These pads have been shown to build contact on smooth surfaces and secrete a lipid-bearing fluid into the contact zone (Peattie et al. 2011). However, it is not known, to which extent the arolia are of importance for attachment or friction generation in these animals. Usually they are too small to generate sufficient friction and adhesive forces, and to permit climbing on smooth surfaces. That is why most solifuges use the pedipalpal adhesive pads (see below) for this purpose (Cushing et al. 2005; Klann 2009, and A. Klann pers. comm.). However, at least one solifuge species (Ammotrichidae: *Oltacola chacoensis*) is able to run on vertical smooth glass panes with remarkable speed (Klann 2009, and A. Klann pers. comm.). This species is relatively small and dwells in bush-like vegetation, in contrast to most other species that are barely observed climbing. The arolium in this species exhibits two large lateral lobes that may be inflatable (Klann 2009). However, the exact function and mechanism of operation has not been studied in vivo and thus remain unclear.

Of much more importance are the arolia of the pedipalps, which are used to climb on the smooth surfaces (Kästner 1933; Cushing et al. 2005), to capture prey (Kästner 1933; Willemart et al. 2011) and to take up water to bring it to the mouth (Kästner 1933). In order to climb smooth vertical surfaces, both pedipalps are alternately attached to the substrate (Cushing et al. 2005). Hence, the whole body weight is supported by one pad alone, which is quite remarkable, as solifuges can reach a body length up to 15 cm and a mass of several grams. Insects flying by or jumping crickets can be retained just by the touch of the pedipalp tip (Kästner 1933; Willemart et al. 2011), which underlines its adhesive strength. Willemart et al. (2011) estimated a median adhesive strength of 7.3×10^4 N/m^2, which is 35 times larger than that of the arolium on the walking leg of a bush cricket (Jiao et al. 2000). The pedipalpal adhesive organ originates from the pretarsus similar to the arolia of the walking legs. However, it might be an independent development due to its rather different structure (Klann 2009). Although the pedipalpal pads have been called 'suctorial organs' (Cushing et al. 2005), it is unlikely that a suction effect is involved in its adhesive mechanism. Ultrastructural analysis revealed that the cuticle of the contacting pad exhibits parallel, apically branched fibrils, very similar to those of smooth adhesive foot pads (Klann et al. 2008). This indicates a comparable functional mechanism. Already Kästner (1933) proposed that the high compliancy of the pad may permit a close contact generation, which, together with a presumed secretion, may produce high adhesive forces. Previous authors argued that secretions may unlikely be involved, because they did not find according glands in the distal pedipalp and pores in the epicuticle of the pad (Cushing et al. 2005; Klann et al. 2008). However, visible pores are not necessarily needed for fluid emergence, like our data on the pseudoscorpion arolium show (see below). Since smooth adhesive pads are proposed to be dependent on the capillarity of a thin fluid film to generate considerable adhesive forces (Jiao et al. 2000; Drechsler and Federle 2006; Barnes 2007), and secretion emergence was observed in the arolia of solifuges (Peattie et al. 2011), the same is very probable for the pedipalpal pad. The pedipalpal pad is usually retracted into the tarsus, apically covered by three lid-like sclerites and activated by inflation, presumably by internal hemolymph pressure (Kästner 1933; Cushing et al. 2005; Klann et al. 2008). For this purpose, the ventral apical sclerite is opened by the attached pretarsal protractor muscle and the dorsal ones presumably by a rise of the internal hemolymph pressure (Klann et al. 2008). Their closing is presumably driven by stored elastic energy of the hinge membranes and the retraction of the attached pad. The pad is retracted by the medially attached pretarsal retractor muscle (Klann et al. 2008).

7.2.4 Mites and Ticks

Smooth adhesive pads are widespread among ticks and mites and inconsistently called pulvillus, sucker, adhesive disc, ambulacral disc or caruncle. These pads arise from the membranes associated with the pretarsus or are a modification (softening) of the entire pretarsus. According to the terminology defined in Chap. 1, we call these organs arolia. Previous authors often presumed a sucker function of these pads (Evans 1992), however, in most cases this is unlikely. The pads presumably function as smooth adhesive pads and are usually fluid mediated, as found in *Androlaelaps schaeferi* (Mesostigmata: Laelapidae) (Peattie et al. 2011), *Ixodes* spp. (Ixodida: Ixodidae) (Baker 1997) (and this study, Fig. 7.3f), and *Anystis* sp. (Trombidiformes: Anystidae) (this study, Fig. 7.9c). In the following, we summarize the distribution, structure and function of the arolia for each of both mite super-orders.

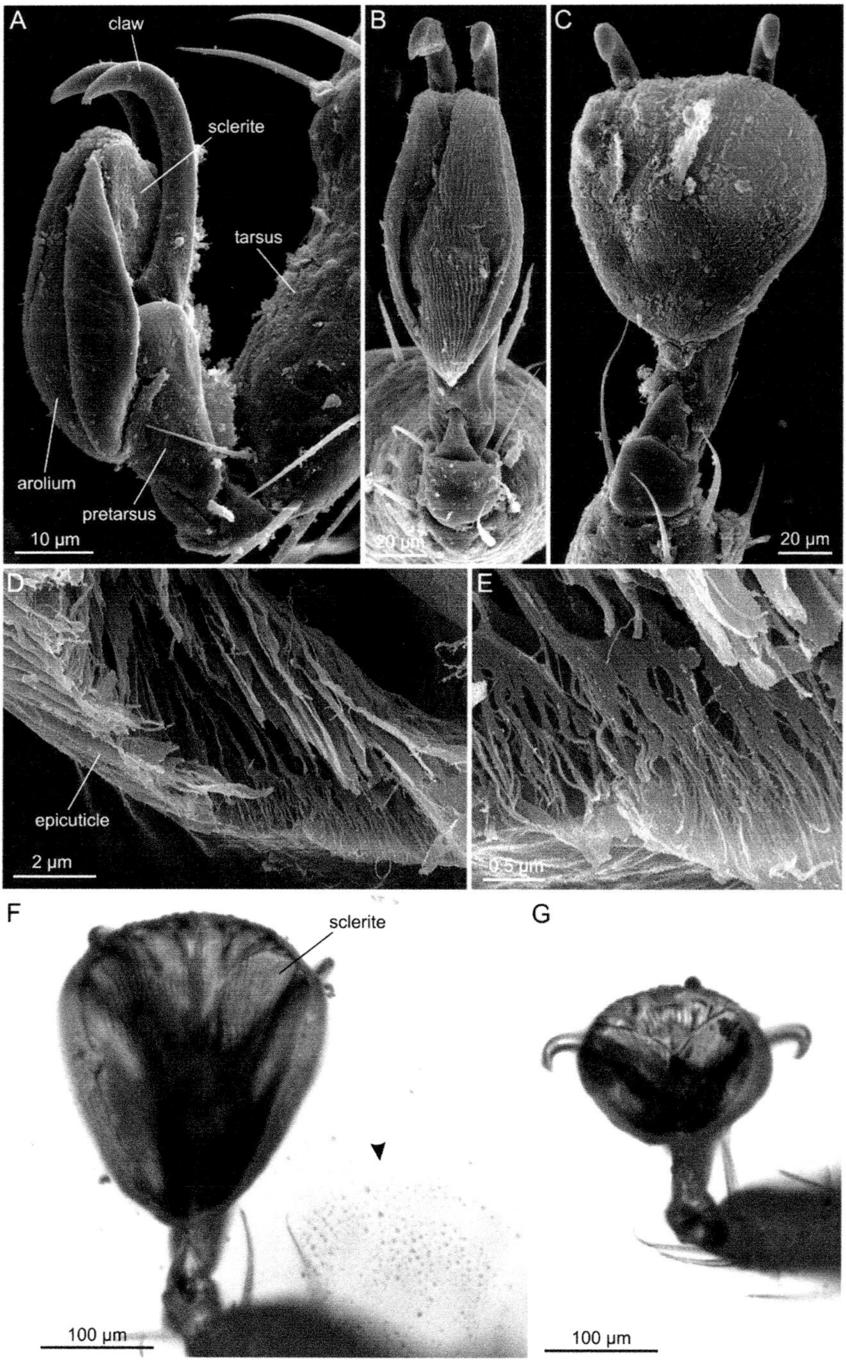

Fig. 7.3 Arolium of Ixodida. Ticks (here: *Ixodes ricinus*) bear large, inflatable arolia used for attachment to plant surfaces for ambushing and to the skin of their vertebrate hosts. (**a**) The lateral view of a hind leg shows the position of the arolium on the pretarsus and the dorso-lateral enclosing sclerites. (**b**) In resting position, the arolium is folded and of a slender oval shape. (**c**) During

Parasitiformes Arolia are found among most parasitiform mites and might be homologous in most cases. The arolia of Opilioacarida and Ixodida are traditionally called *'pulvilli'*, however, we do not use this term since the pads obviously arise from the median pretarsus.

In the Opilioacarida, the arolium is expandable and pad-like (van der Hammen 1966; Alberti and Coons 1999). In the first leg pair, which is feeler-like elongated, the arolium is lacking (pers. observation).

In ticks (Ixodida), all stages possess arolia in the Ixodidae (Baker 1997) and Nuttalliellidae (Latif et al. 2012). However, in the Argasidae, they are only present in the larvae (van der Hammen 1983b). The arolium is very similar between these families. In *Ixodes* spp., the arolium is a large inflatable pad, with a pair of dorsal sclerites that enclose the pad in the retracted state (Baker 1997) (Fig. 7.3a-c; Video 7.8). The inflation is likely driven by an internal rise of hemolymph pressure, and the retraction by the pulling of the pretarsal retractor muscle and the elasticity of the closing dorsal sclerites. The arolia in the first leg pair are much larger than those of the other legs (Fig. 7.3f). The inner cuticle of the contacting pad forms thin walled hollow tubes, which apically disintegrate into branched fibers (Figs. 7.3d, e). The enclosing epicuticle is very thin and, due to the underlying structure, highly compliable. In between there is a lipid-bearing fluid, which also emerges into the contact through pores in the epicuticle, leaving discernable footprints when walking on a glass slide (Baker 1997) (Fig. 7.3f).

In the Holothyrida the arolium is of a shape comparable to that of Ixodida, but it is lacking in the first leg pair (van der Hammen 1983a).

The diverse Mesostigmata contain mites with different ecology, including predatory and parasitic mites. Pad-like inflatable arolia are widespread. Usually the whole distal section of the ambulacrum is membraneous, except for the, usually small, dorsally attached pair of claws (Alberti and Coons 1999). In some specialized parasites the claws may totally lack, as in the bee parasite *Varroa* spp. (Akimov et al. 1988) (Fig. 7.4c). In many free living mesostigmatids the basal ambulacrum forms a long tube-like stalk, in which the arolium can partly be retracted (Evans and Till 1965) (Fig. 7.4d). The expanded arolium may form a small median and a pair of large apical lobes (Fig. 7.4e).

Acariformes Among Trombidiformes, there are lineages with hairy adhesive pads (see Chap. 3), smooth adhesive pads (Heterostigmata), with both types (Anystinae), or those lacking both and having enlarged claws instead (e.g. Hydracarina, Halacaroidea). In the Anystidae, the arolium is only present in the sub-family Anystinae (Meyer and Ueckermann 1987). It is presumably derived from the plesiomorphic empodium, like the hairy pads in the Eupodina. The arolium is very similar to that of pseudoscorpions (see Sect. 7.4 for more details). In many Heterostigmata

Fig. 7.3 (continued) pretarsal depression, the arolium gets inflated, thereby highly expanding in lateral directions. (**d**, **e**) The ventral cuticle is thickened and soft, with hollow tubes, disintegrating into fine branches towards the epicuticle, as seen in this fracture. (**f**, **g**) Arolia in contact with a smooth glass surface (viewed from below); (**f**) in the first leg pair the arolia are much larger than in the other legs (compare with leg 4 in (**g**)), *arrowhead* indicates a footprint of tiny lipid-bearing droplets

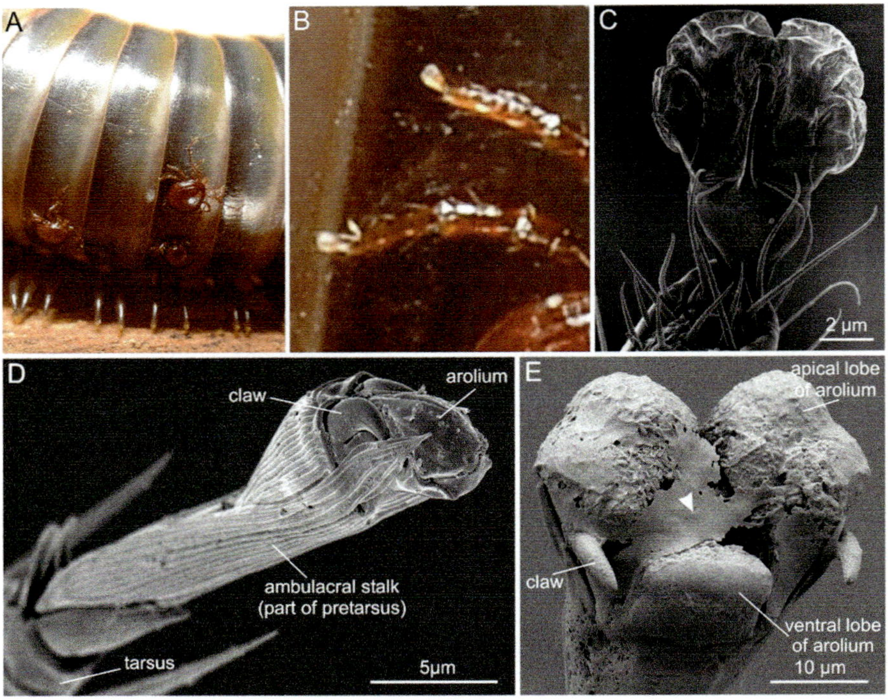

Fig. 7.4 Arolium of Mesostigmata. (**a**, **b**) The mite *Scissuralaelaps huberi* (Laelapidae) is associated with millipedes (Seeman and Alberti 2015); it attaches to the smooth cuticle of the host by large arolia (see (**b**) as a detail of (**a**)) (photos by S. Huber, with kind permission). (**c**) The bee parasite *Varroa jacobsoni* exhibits large disc-like arolia, the claws are totally lacking (dorsal view) (adapted from Alberti and Coons 1999, with kind permission by G. Alberti and Wiley). (**d**) The arolium of mesostigmatid mites can be retracted into the elongated pretarsus (ambulacral stalk) (here: *Amblyseius swirskii* (Phytoseiidae), distal portion of leg 4, lateral-apical view). (**e**) It may be inflated by hemolymph pressure (here: Cryo-SEM of leg 3 of an unidentified mesostigmatid, frontal-ventral view); note the presence of a secretion (*arrowhead*)

(e.g. Tarsonemidae and Heterocheylidae), the basal articulation membrane of the single claw is highly inflated and forms a large arolium, which is used to attach to hosts (Lindquist 1986). In some families, the arolium is built by a modified empodium and both claws are still present, e.g. in Pygmephoridae (Khaustov and Ermilov 2011).

In the Sarcoptiformes, the large clade of Astigmata includes many parasitic mites that have evolved several types of arolia (Atyeo 1979; OConnor 1982; Alberti and Coons 1999; Krantz and Walter 2009). Often the whole distal section of the ambulacrum forms a soft cushion or large disc (often called *'caruncle'*, *'ambulacral disc'* or *'sucker'*) with a small, embedded single claw and dorsal sclerites, both of which can also be totally reduced. In the Oribatida, adhesive pads are very rare, but some arboreal species exhibit a pad like extension of the ventral claw articulation membrane (called *'pulvillus'*), i.e. in *Adhaesozetes polyphyllos* (Adhaesozetidae) (Walter and Behan-Pelletier 1993), *Phylleremus* spp. (Licneremaeoidea) (Behan-Pelletier and Walter 2007), *Symbioribates aokii* (Oripodoidea) (Karasawa and Behan-Pelletier 2007) or *Zetorchestes falzonii* (Zetorchestidae) (Alberti and Coons 1999). In these cases, the size of the arolium increases from front to hind legs, whereas the size of the single claw decreases, which may suggest a division of labor between different legs in attachment to different surfaces. However, to which extent the rather small arolium generates adhesion has not been studied. For Zetochestidae, which are capable of leaps, Alberti and Coons (1999) argue that the arolium may help to generate the necessary thrust to take off from smooth leaf surfaces.

7.2.5 Spiders

Many spiders exhibit hairy adhesive pads (see Chap. 3), but there is only a single case of a non-sclerotized pad-like protrusion in the pretarsus, which was recently found in *Nopsides* spp. (Caponiidae) and called *'pulvillus'* by the authors (Jiménez et al. 2011). According to our definitions this structure is an arolium and is likely an adhesive pad, although its actual function is completely unknown. Another, much earlier, report of an arolium in the front legs of a jumping spider (De Meijere 1901) is likely an artefact, since such structure has never been found in salticids by other authors.

7.3 Disc-Like Arolium

In some arachnids, the arolium is flattened and disc-like. In such cases, the claws are often totally lacking. The tarsus may be attached to a median part, hence these arolia may also (to some extent) work in the manner similar to that of a mushroom-like microstructure (see Chap. 5).

One example of such arolium is known from the feet of whip-scorpionprenymphs. Like in scorpions, whip-scorpion mothers carry their hatched prenymphs on their abdomen. Proper attachment of the prenymphs relies on their large disc-like arolia (Kästner 1941; Wolff et al. 2015a). In the Thelyphonida, the prenymph pretarsus lacks all sclerotized structures including the claws and consists of the large arolium only (Fig. 7.5). The prenymph leg lacks clear segmentation and is barely moveable. The arolium sticks passively due to the high compliance of the ventral cuticle and the viscous and capillary forces of a viscous secretion (Wolff et al. 2015a) (Fig. 7.5g). Hence, the previous description of this organ being a sucker (Kästner 1941) is erroneous. The high compliance of the ventral cuticle is due to its sponge-like structure with arrays of thin, branched fibers underneath the thin, but robust epicuticle (Fig. 7.5c, d). In contrast, the dorsal cuticle is rather stiff, with a much denser material structure, and forms denticulate protuberances, which further reduce the potential contact area (Fig. 7.5b, c). In the Schizomida, both morphology and

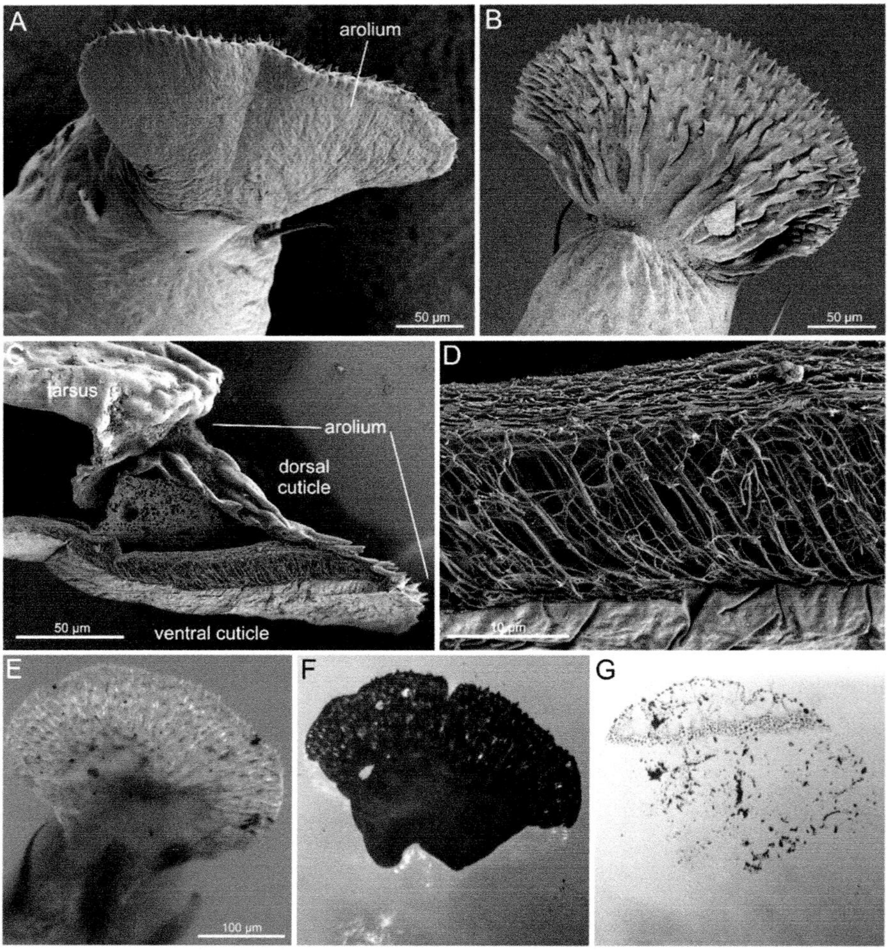

Fig. 7.5 Arolium of whip-scorpion prenymphs. In the prenymphs of whip-scorpions (here: *Typopeltis crucifer*), the whole pretarsus is modified into a large disc-like arolium. (**a**) The ventral side is smooth, as seen in this Cryo-SEM image, and highly sticky. (**b**) The denticulated dorsal side is non-adhesive. (**c**) The section reveals the totally different structure, thickness and density of dorsal and ventral cuticle. (**d**) The ventral cuticle is highly thickened and the material is disintegrated into fibrils, making this part of the pad highly compliant. (**e–g**) An arolium of a freshly ablated leg sticks passively to a smooth glass slide (viewed from below); (**f**) the whole portion of the ventral side generates the contact (visualized by coaxial illumination) and (**g**) viscous secretion is left behind on the glass surface after detachment

function of this organ is not known, however, it is presumed to be comparable with that of thelyphonid prenymphs due to the close relationship and the similar pulli-carrying behavior.

In some parasitic mites, the whole pretarsus is modified into a disc-like adhesive pad. For instance, the common bee parasite*Varroa* spp. (Mesostigmata) exhibits large disc-like arolia (*'caruncle'*) (Fig. 7.4c), which also can be fully retracted into the ambulacral stalk (Akimov et al. 1988; Liu and Peng 1990). However, in adult females the pad bears claw-like sclerites, which were proposed to be used to grasp the setae of bee hosts (Ramirez and Malavasi 1991). This shift from a smooth pad to an interlocking device reflects the adaptation to a change in the mode of life: The immature stages primarily move on the smooth honeycomb walls of the bee's nest, whereas the adult females are phoretic in order to enhance their dispersal ability (Ramirez and Malavasi 1991). Many of the parasitic astigmatid mites bear large disc-like arolia (*'ambulacral discs'*) (Atyeo 1979; OConnor 1982; Alberti and Coons 1999; Krantz and Walter 2009). From the contact mechanics point of view, such discs may represent mushroom-like contact elements, since the ambulacral stalk is dorso-medially attached.

7.4 Transverse Lip Arolium

In some arachnids, the arolium is distally broadened, such that the overall structure is triangular. Only the distal edge, we call it 'lip', is capable of contact generation with the substrate. In this part, the cuticle is extremely compliant, whereas the basal part of the arolium, the 'trunc', exhibits a stiff, rigid cuticle, keeping the shape of the arolium. This specific type of arolium seems to be very efficient, since it has been evolved at least three times independently in arachnids.

7.4.1 Whip-Spiders

In whip-spiders (Amblypygi), the first instar emerging from the egg, the prenymph, is not yet fully developed and stays on the mother, until next molding. For the attachment onto the mother, the prenymph bears lobe-like adhesive pads between the claws (Quintero 1975; Wolff et al. 2015a, b). The basal lineages of whip spiders, including the Paracharontidae, Charontidae, and Charinidae, retain the pad in the subsequent instars (Weygoldt 2000). The arolium in representatives of these groups has an elaborate structure, combining features of smooth and hairy adhesive pads (see also Chap. 3). The contact is made by a distal transverse-oval lip, in which the cuticle is bloated and consists of numerous parallel fibers that are branching for multiple times towards the epicuticle (Fig. 7.6c). In the distal part, the fine branches merge with the epicuticle, which forms hollow hexagonal tubes that are apically keeled and equipped with spatula-like tips (Fig. 7.6c–e). This is a unique feature of the whip-spider arolium, having large effects on the contact mechanics and attachment strength (see Chap. 3) (Video 7.11). The branching fibers originate from parallel rigid ribs in the endo-cuticle. Their orientation stiffens the pad in a longitudinal direction, while permitting high bending in the transverse direction. This

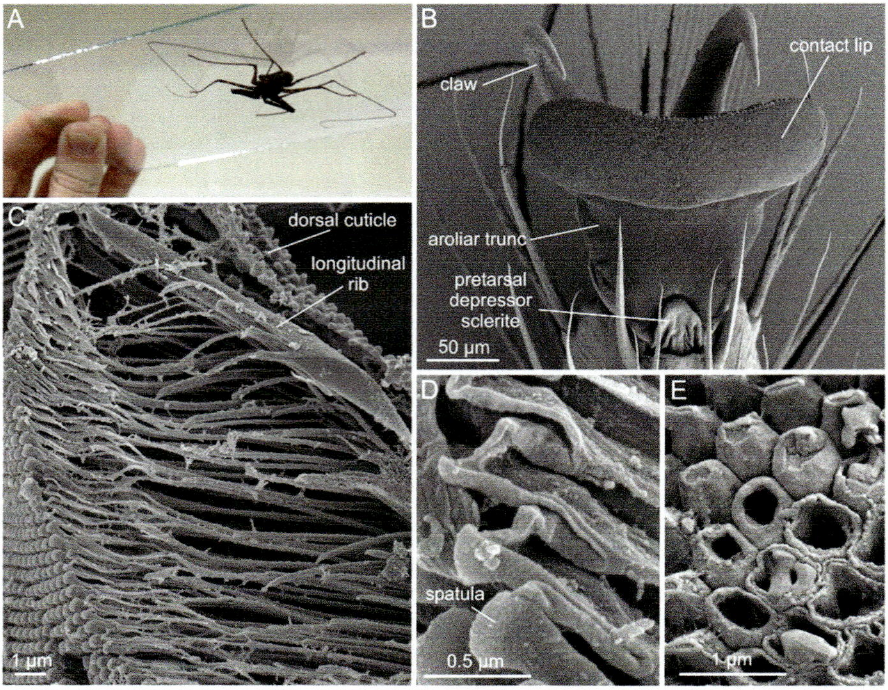

Fig. 7.6 Arolium of whip-spiders. (**a**) Even large whip-spiders of the Charontidae (here: *Charon* cf. *grayi*) can easily walk upside-down on glass slides. (**b**) Their six walking legs bear arolia consisting of a rather stiff trunk and a laterally broadened lip, which makes the contact and exhibits a specific microstructure (here: leg 3 of a *Charon cf. grayi* protonymph). (**c-e**) Sectioning reveals the cuticular structure of the lip region (here: hind legs of *Charinus cubensis* (Charinidae), (**c**, **d**) longitudinal sections, (**e**) transverse section); (**c**) the ventral (contacting) cuticle consists of stiff longitudinal ribs, from which parallel fibers branch off, further splitting into fibrils in the distal fourth; in comparison the dorsal cuticle is very thin and dense, it has a nubby surface and is foldable; (**d**, **e**) the ventral epicuticle forms hexagonal tubes with an apical spatulate keel

configuration permits a mechanism of switching between rather uniform stress distribution (high adhesion) and local stress concentration (quick, effortless detachment).

The amblypygid arolium is working both in a dry and a wet mode. In the first case, no fluid secretions are visible and no traces are left behind on the glass substrate after detachment (Video 7.11). In the second case, a fluid secretion emerges in between the spatulae and droplets are left behind after detachment (Video 7.12). In most cases, only parts of the arolium are wetted (Video 7.13). The occasional secretion may vary enormously in its composition and viscosity. This is indicated by some footprints forming droplets with the major fraction consisting of volatile compounds, while other footprints remain exactly in their original shape (Video 7.14), with the imprints of spatulae visible in the scanning electron microscope (Wolff et al. 2015b).

7.4.2 Pseudoscorpions

Underneath the claws, pseudoscorpions bear an arolium. To the best of our knowledge, no pseudoscorpion species completely lacking the arolium has ever been described. The arolium is already present in the fossil *Dracochela deprehendor* from Devonian, which has been considered to belong to a stem-group of pseudoscorpions (Judson 2012). The arolium has formerly been interpreted as a funnel-like suction cup (Beier 1932), which is erroneous. Our observations show that no cavity is present in the contact (Fig. 7.8). The arolium has a narrow base and widens distally to a broad lip (Fig. 7.7a, b, g, k). All parts except the distal lip exhibit a rigid cuticle, which, however, is not tanned and melanised, but rather translucent, and may exhibit a microstructured surface. In the proximal and lateral parts, the cuticle is comparably thick (1–3 μm) and dense, whereas in the median dorsal part it is thinner, as observed in histological sections.

The pretarsal retractor tendon extends into the arolium and is attached to its dorsal median part. The lip cuticle is very thin and hence highly compliant (Fig. 7.7c, d, l). The underlying material may be fibrous. However, this could not clearly be seen in the sections, because a dense fluid was present in the inner cavity of the arolium, which was fixed during sample preparation (Fig. 7.7d). In *Neobisium*, the lip cuticle externally exhibits nubs (Fig. 7.7c), which might be the bulges of underlying fibers. The lip may be partly folded at the lateral sides and expanded by an increase of hemolymph pressure. In contrast, the arolia of the Cheliferoidea (Fig. 7.7i–k) and Garypoidea (Fig. 7.7g, h) seem to be rather stiff and stable in shape, with the exception of the distal lip. The arolium is usually slightly shorter than the claws, but can also be much longer, as in many Olpiidae and Garypiidae (Fig. 7.7g, h) (Beier 1932; Harvey 1992). In the Garypinidae, the arolium is divided into two halves (Harvey and Šťáhlavský 2010). In some genera of Ideoroncidae, the arolium bears a median hook-like protuberance (Harvey and Mahnert 2006). Whether it is stiff or soft and how it affects substrate attachment remains unknown. In the Parahyidae, the arolium laterally bears two pointed protuberances (Harvey 1992).

The contact of the arolium is only made by the distal lip (Fig. 7.8b). The contact area resembles a transverse, slightly convex, narrow strip. During detachment, the delamination zone propagates from lateral to median (Fig. 7.8c, d; Videos 7.15 and 7.16), like in the whip-spider arolium. This is presumably facilitated by the higher rigidity of the lateral parts, which better transmit the tensile stress, when the pretarsal retractor contracts, whereas the more flexible median section partly absorbs stress and stays longer in contact. Adhesive fluids are secreted into the contact and a thin film may be left behind on the substrate after detachment (Fig. 7.8d). These footprints shrink to droplets on the (hydrophilic) glass and do not evaporate, which indicates their lipid-like nature. Since no pores were visible in the lip cuticle, the fluids probably diffuse through it, which may only be possible if the fluid has a low viscosity.

During pull-off tests, *Neobisium* sp. individuals could resist tensile forces of 100–700 μN, which corresponds to a safety factor of 8–45 times of their body weight.

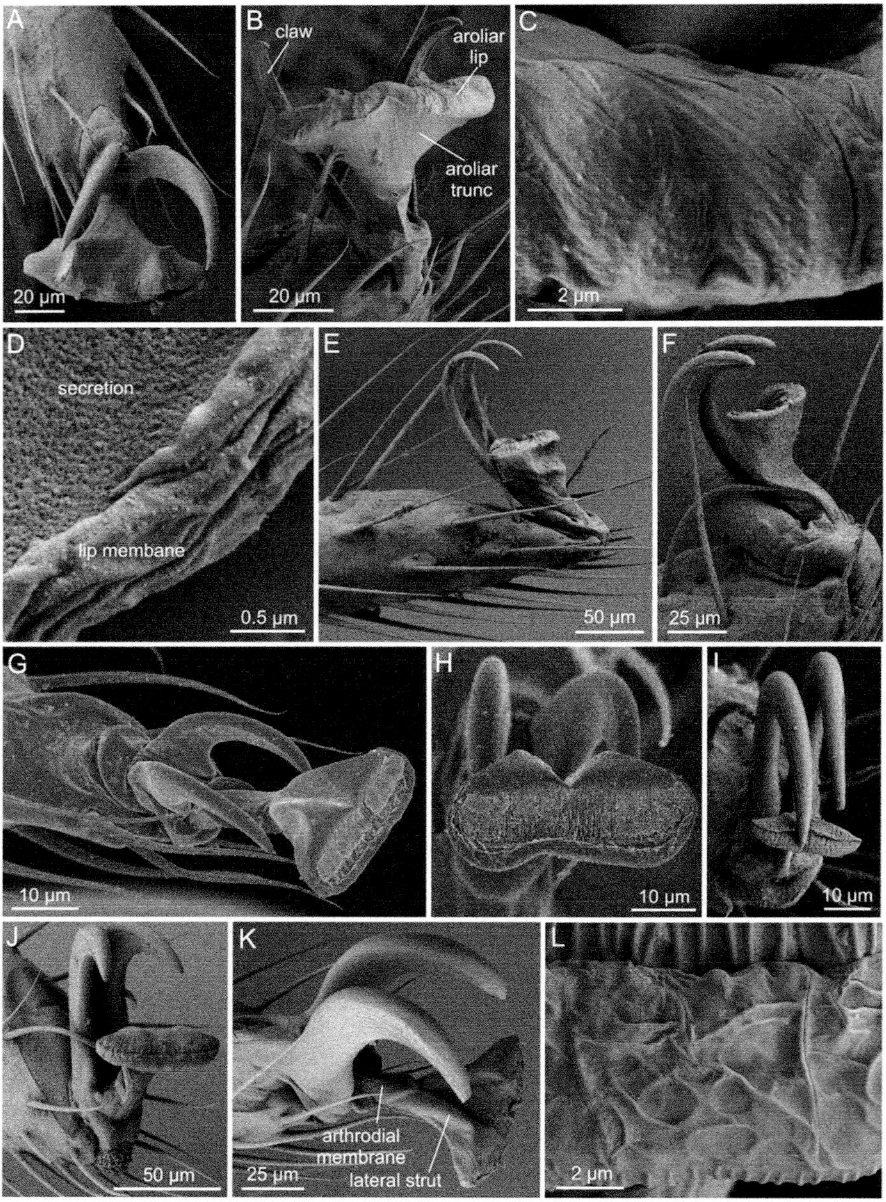

Fig. 7.7 Arolium of pseudoscorpions. All pseudoscorpions possess arolia in all four walking legs, consisting of a stiff trunk and a broad contact lip. (**a–d**) *Neobisium* sp. (Neobisiidae), Cryo-SEM (except (**d**)); (**a**) leg 2, pretarsus protracted, dorsal view; (**b**) same, pretarsus partly retracted, ventral view; (**c**) detail of aroliar lip, frontal view; note the nubby surface, and the minute folds indicating extreme thinness and high compliancy of the outer membrane; (**d**) section of aroliar lip; the interior of the arolium is filled with a dense viscous fluid; the outer membrane consists of the epicuticle only. (**e**) *Neobisium spelaeum*, leg 2, lateral view. (**f**) *Megachernes himalayensis* (Chernetidae), leg 4, lateral view. (**g**, **h**) Garypidae, unidentified species, note the elongated trunk, shifting the lip away from the claw tips; (**g**) leg 2, dorsal-lateral view; (**h**) leg 4, frontal view. (**i**) *Pselaphochernes scorpioides* (Chernetidae), leg 4, frontal view. (**j–l**) *Atemnus* sp. (Atemnidae), Cryo-SEM; (**j**) leg 2, frontal view; (**k**) leg 3, dorsal-lateral view; (**l**) detail of aroliar lip, frontal view

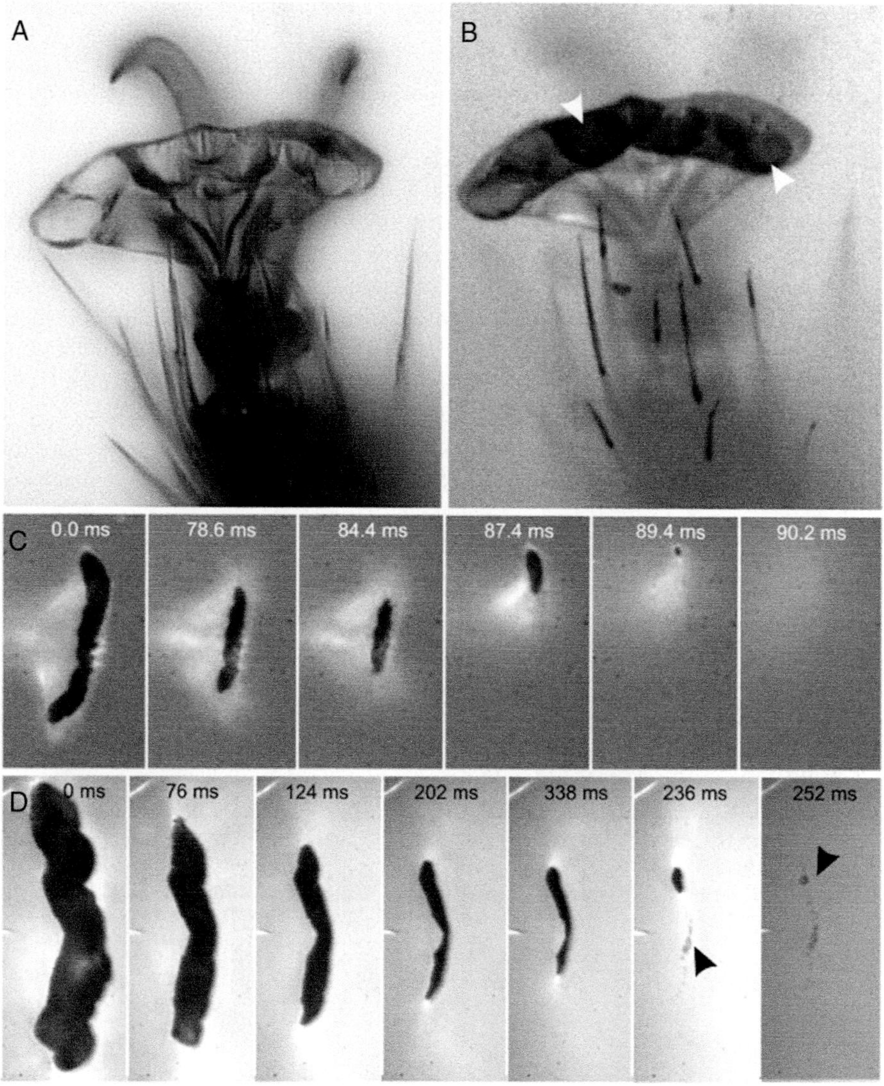

Fig. 7.8 Contact mechanics of the pseudoscorpion arolium. (**a**, **b**) *Neobisium* sp. (Neobisidae). Light micrographs of the pretarsus of a pseudoscorpion walking upside-down on a glass slide, as viewed through the slide; (**a**) transmission light micrograph; note the two lateral struts embedded in the aroliar trunk; (**b**) same, with coaxial lighting; the contact area appears dark; note the local presence of secretion in the contact, indicated by a darker contrast (*arrowheads*). (**c**) Frames of a RICM-HSV sequence (frame rate 5000 fps) of the detaching arolium of leg 4 of a *Chthonius* sp. (Chthoniidae) walking on a glass cover slip; note the lateral peel-off of the pad. See Video 7.15 for motion. (**d**) Frames of a RICM-HSV sequence (500 fps, shutter speed 1/5000) of the detaching arolium of leg 4 of a *Neobisium* sp. walking on a glass cover slip; note the minute droplets of secretion remaining after detachment (*arrowheads*). See Video 7.16

7.4.3 Anystid Mites

An arolium very similar to that of pseudoscorpions is present in trombidiform mites of the subfamily Anystinae. The arolium shape is triangular. It is highly broadened distally and consists of rather stiff cuticle, except for the distal most edge (Fig. 7.9a). Hence, the contact area resembles a narrow transverse strip (Fig. 7.9b). The arolium adheres using capillary forces of a lipid-like secretion (Fig. 7.9c; Video 7.19), but fluids are not always left on the substrate after detachment (Video 7.20). Interestingly, these mites additionally bear a pair of large, branched tenent setae, which further support adhesion. The contact area of the arolium is located more distally than that of the tenent setae (Fig. 7.9b). During attachment, the tenent setae make first contact, then the arolium, and during detachment vice versa (Video 7.19). Occasionally only the setae make the contact (Video 7.21).

Hence, the tenent setae may primarily serve friction generation for securing the steps and generating forward thrust, whereas the arolium may securely attach the

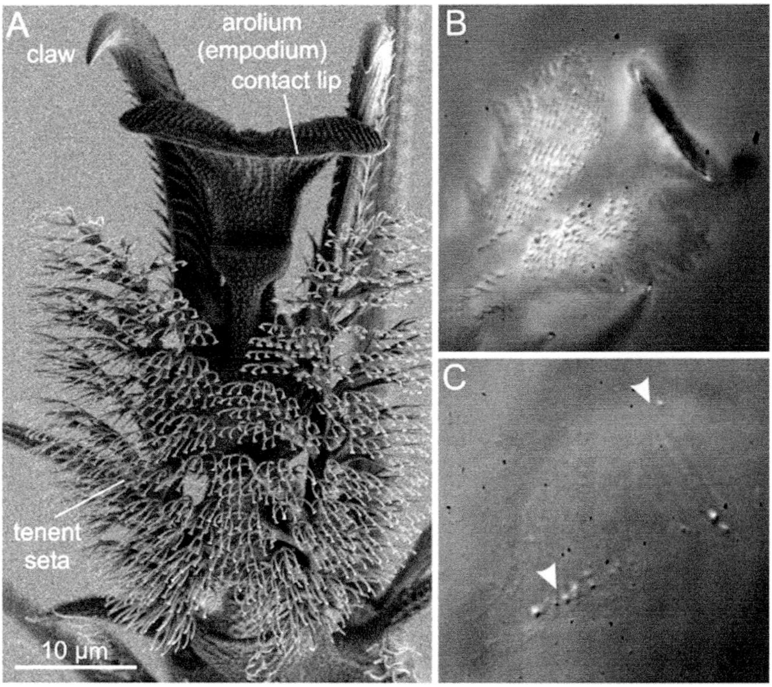

Fig. 7.9 Attachment devices of mites from the sub-family Anystinae. (**a**) In the Anystinae (here *Anystis* sp.), the empodium is modified into a transverse lip-shaped arolium, similar to that of pseudoscorpions; additionally, there is a pair of large, branched tenent setae (see Chapt. 3 for details on that). (**b**) The RICM image of a hind leg in contact to the glass slide shows the narrow, transverse contact area of the lip (*dark area*). (**c**) The RICM image reveals droplets of a lipid-bearing secretion left behind from both the arolium and the tenent setae (*arrowheads*) after foot detachment

animal between the steps. *Anystis* sp. mites can run very fast: 42 ± 14 body lengths per second (mean ± s.d., N = 16). This implicates that the contacts must be generated and loosened very quickly, which is quite remarkable.

7.5 Other Smooth Adhesive Pads

In some instances, smooth adhesive pads may occur in other body parts than feet, like in the anal region of Heterozerconidae mites (Mesostigmata). These have been previously described as suckers (Baker et al. 1987), however, they do not exhibit an inner cavity, and hence we interpret these as smooth adhesive pads. These pads are found in the parasitic adults that use them to attach to hosts, like millipedes (Farfan and Klompen 2012). They consist of two soft protuberances situated laterally to the anal opening and can be retracted by muscles (Baker et al. 1987). Details on their ultrastructure and on the possible presence of secretions are lacking. Another anal pad is the sucker of Eriophoridae (see Chap. 6), which may also function as a smooth adhesive pad.

References

Akimov IA, Starovir IA, Yastrebtsov AV, Gorgol VG (1988) Varroa mite: the agent of the Bee varroatosis. Naukova Dumka, Kiev (in Russian)

Alberti G, Coons L (1999) Acari: mites. In: Harrison FW, Foelix RF (eds) Microscopic anatomy of invertebrates: chelicerate Arthropoda, vol 8. Wiley-Liss, New York, pp 515–1265

Atyeo WT (1979) Pretarsi of astigmatid mites. Acarologia 20:244–269

Baker GT (1997) The pulvillus: Cuticular structure and function (Acarina: Ixodida). 日本ダニ学会誌 6(1):25–31

Baker G, Chandrapatya A, Nesbitt H (1987) Morphology of several types of cuticular suckers on mites (Arachnida, Acarina). Spixiana 10:131–137

Barnes WJ (2007) Functional morphology and design constraints of smooth adhesive pads. MRS Bull 32(06):479–485

Behan-Pelletier VM, Walter DE (2007) *Phylleremus* n. gen., from leaves of deciduous trees in eastern Australia (Oribatida: Licneremaeoidea). Zootaxa 1386:1–17

Beier M (1932) 5. Ordnung der Arachnida: Pseudoscorpionidea = Afterscorpione. In: Kükenthal W, Krumbach T (eds) Handbuch der Zoologie: Bd. 3, 2. Hälfte. Chelicerata, vol 3. W. de Gruyter & Co, Berlin, pp 117–192

Bennemann M, Backhaus S, Scholz I, Park D, Mayer J, Baumgartner W (2014) Determination of the Young's modulus of the epicuticle of the smooth adhesive organs of *Carausius morosus* using tensile testing. J Exp Biol 217(20):3677–3687

Bullock JM, Drechsler P, Federle W (2008) Comparison of smooth and hairy attachment pads in insects: friction, adhesion and mechanisms for direction-dependence. J Exp Biol 211(20):3333–3343

Cushing PE, Brookhart JO, Kleebe H-J, Zito G, Payne P (2005) The suctorial organ of the Solifugae (Arachnida, Solifugae). Arthropod Struct Dev 34(4):397–406

De Meijere JC (1901) Über das letzte Glied der Beine bei den Arthropoden. Zool Jahrb Anat 14:417–476

Drechsler P, Federle W (2006) Biomechanics of smooth adhesive pads in insects: influence of tarsal secretion on attachment performance. J Comp Physiol A 192(11):1213–1222
Evans GO (1992) Principles of acarology, vol 18. CAB international, Wallingford
Evans GO, Till WM (1965) Studies on the British Dermanyssidae (Acari: Mesostigmata). Part I. External morphology. Bull Br Mus(Nat Hist) Zool 13(8):249–294
Farfan M, Klompen H (2012) Phoretic mite associates of millipedes (Diplopoda, Julidae) in the northern Atlantic region (North America, Europe). Int J Myriapodol 7:69–91
Federle W, Endlein T (2004) Locomotion and adhesion: dynamic control of adhesive surface contact in ants. Arthropod Struct Dev 33(1):67–75
Gnaspini P (2007) Development. In: Pinto-da-Rocha R, Machado G, Giribet G (eds) Harvestmen: the biology of Opiliones. Harvard University Press, Cambridge
Gnaspini P, da Silva MB, Pioker FC (2004) The occurrence of two adult instars among Grassatores (Arachnida: Opiliones) – a new type of life-cycle in arachnids. Invertebr Reprod Dev 45(1):29–39
Gorb S (2000) Biological microtribology: anisotropy in frictional forces of orthopteran attachment pads reflects the ultrastructure of a highly deformable material. Proc R Soc B 267(1449):1239–1244
Gorb S (2001) Attachment devices of insect cuticle. Springer Science & Business Media, Dordrecht
Gorb SN (2007) Smooth attachment devices in insects: functional morphology and biomechanics. Adv Insect Physiol 34:81–115
Gorb SN, Beutel RG, Gorb EV, Jiao Y, Kastner V, Niederegger S, Popov VL, Scherge M, Schwarz U, Vötsch W (2002) Structural design and biomechanics of friction-based releasable attachment devices in insects. Integr Comp Biol 42(6):1127–1139
Harvey M (1992) The phylogeny and classification of the Pseudoscorpionida (Chelicerata: Arachnida). Invertebr Syst 6(6):1373–1435
Harvey MS, Mahnert V (2006) The systematic position of the Amazonian species of *Albiorix* (Pseudoscorpiones, Ideoroncidae). J Arachnol 34(1):227–230
Harvey MS, Šťáhlavský F (2010) A review of the pseudoscorpion genus *Oreolpium* (Pseudoscorpiones: Garypinidae), with remarks on the composition of the Garypinidae and on pseudoscorpions with bipolar distributions. J Arachnol 38(2):294–308
Jiao Y, Gorb S, Scherge M (2000) Adhesion measured on the attachment pads of *Tettigonia viridissima* (Orthoptera, Insecta). J Exp Biol 203(12):1887–1895
Jiménez ML, Platnick NI, Dupérré N (2011) The haplogyne spider genus *Nopsides* (Araneae, Caponiidae), with notes on *Amrishoonops*. Am Mus Novit 3708:1–18
Juberthie C (1972) Reproduction et developpement d'un opilion Cosmetidae, *Cynorta cubana* (Banks), de Cuba. Ann Spéléol 27:773–785
Judson ML (2012) Reinterpretation of *Dracochela deprehendor* (Arachnida: Pseudoscorpiones) as a stem-group pseudoscorpion. Palaeontology 55(2):261–283
Karasawa S, Behan-Pelletier V (2007) Description of a sexually dimorphic oribatid mite (Arachnida: Acari: Oribatida) from canopy habitats of the Ryukyu Archipelago, southwestern Japan. Zool Sci 24(10):1051–1058
Kästner A (1933) 6. Ordnung der Arachnida. Solifugae Sundvall Walzenspinnen. In: Kükenthal W, Krumbach T (eds) Handbuch der Zoologie, vol 3, vol 2. de Gruyter & Co., Berlin, pp 193–299
Kästner A (1941) 2. Ordnung der Arachnida: Pedipalpi Latreille = Geißel-Scorpione. In: Kükenthal W, Krumbach T (eds) Handbuch der Zoologie: chelicerata, vol 3. Walter de Gruyter & Co., Berlin, pp 1–76
Khaustov A, Ermilov S (2011) A new species of the genus *Siteroptes* (Acari, Heterostigmata, Pygmephoridae) from European Russia. Ent Rev 91(4):528–532
Klann AE (2009) Histology and ultrastructure of solifuges. Comparative studies of organ systems of solifuges (Arachnida, Solifugae) with special focus on functional analyses and phylogenetic interpretations. PhD-thesis, Ernst-Moritz-Arndt-University Greifswald, Greifswald. Published

online: https://www.deutsche-digitale-bibliothek.de/binary/ODU3XEEXCIZGVHOOHIRD7XTONYTTEVS7/full/1.pdf

Klann A, Gromov A, Cushing P, Peretti A, Alberti G (2008) The anatomy and ultrastructure of the suctorial organ of Solifugae (Arachnida). Arthropod Struct Dev 37(1):3–12

Krantz GW, Walter DE (2009) A manual of acarology, 3rd edn. Texas Tech University Press, Lubbock

Latif AA, Putterill JF, De Klerk DG, Pienaar R, Mans BJ (2012) *Nuttalliella namaqua* (Ixodoidea: Nuttalliellidae): first description of the male, immature stages and re-description of the female. Plos One 7(7):e41651

Lindquist EE (1986) The world genera of Tarsonemidae (Acari: Heterostigmata): a morphological, phylogenetic, and systematic revision, with a reclassification of family-group taxa in the Heterostigmata. Memoirs Ent Soc Can 118(S136):1–517

Liu TP, Peng Y-SC (1990) Scanning electron microscopic observation of the pretarsal suckers of the honey-bee ectoparasite, *Varroa jacobsoni* (Gamasida: Dermanyssina). Exp Appl Acarol 8(1–2):105–114

Meyer MKPS, Ueckermann EA (1987) A taxonomic study of some Anystidae (Acari: Prostigmata). Entomol Mem Dep Agric Wat Supply Repub S Afr 68:1–37

Millot J, Vachon M (1949) Ordre de Scorpions. In: Grassé P-P (ed) Traité de Zoologie. Tome VI. Masson et Cie, Paris, pp 386–436

OConnor B (1982) Evolutionary ecology of astigmatid mites. Annu Rev Entomol 27(1):385–409

Peattie AM, Dirks J-H, Henriques S, Federle W (2011) Arachnids secrete a fluid over their adhesive pads. Plos One 6(5):e20485

Perez Goodwyn PJ, Peressadko A, Schwarz H, Kastner V, Gorb SN (2006) Material structure, stiffness, and adhesion: why attachment pads of the grasshopper (*Tettigonia viridissima*) adhere more strongly than those of the locust (*Locusta migratoria*) (Insecta: Orthoptera). J Comp Physiol A 192:1233–1243

Quintero D (1975) Scanning electron microscope observations on the tarsi of the legs of Amblypygids (Arachnida Amblypygi). In: Proceedings of the 6th International Arachnological Congress, pp 161–163

Ramirez WB, Malavasi JG (1991) Conformation of the ambulacrum of *Varroa jacobsoni* Oudemans (Mesostigmata: Varroidae): a grasping structure. Int J Acarol 17(3):169–173

Roewer CF (1935) Opiliones (Fünfte série). Zugleich eine Revision aller bisher bekannten europäischen Laniatores. Bioespeleologica LXII Archives de Zoologie Experimental et Géneral 78(1):1–96

Schargott M, Popov VL, Gorb SN (2006) Spring model of biological attachment pads. J Theor Biol 243:48–53

Seeman OD, Alberti G (2015) A new species of *Scissuralaelaps* (Acari: Mesostigmata: Laelapidae) from millipedes in the Philippines. Syst Appl Acarol 20(6):707–720

Sharma PP, Giribet G (2011) The evolutionary and biogeographic history of the armoured harvestmen–Laniatores phylogeny based on ten molecular markers, with the description of two new families of Opiliones (Arachnida). Invertebr Syst 25(2):106–142

Townsend VR, Rana NJ, Proud DN, Moore MK, Rock P, Felgenhauer BE (2009) Morphological changes during postembryonic development in two species of Neotropical harvestmen (Opiliones, Laniatores, Cranaidae). J Morphol 270(9):1055–1068

van der Hammen L (1966) Studies on Opilioacarida (Arachnida) I. Description of *Opilioacarus texanus* (Chamberlin & Mulaik) and revised classification of the genera. Zoologische Verhandelingen 86(1):1–80

van der Hammen L (1983a) New notes on Holothyrida (Anactinotrichid mites). Zoologische Verhandelingen 207:1–48

van der Hammen L (1983b) Notes on the comparative morphology of ticks (Anactinotrichida: Ixodida). Zoologische Mededelingen 57:209–242

Walter DE, Behan-Pelletier VM (1993) Systematics and ecology of *Adhaesozetes polyphyllos* sp. nov. (Acari: Oribatida: Licneremaeoidea), a leaf-inhabiting mite from Australian rainforests. Can J Zool 71(5):1024–1040

Weygoldt P (2000) Whip spiders (Chelicerata: Amblypygi): their biology, morphology and systematics. Apollo Books, Stenstrup, 163 pp

Willemart RH, Santer RD, Spence AJ, Hebets EA (2011) A sticky situation: solifugids (Arachnida, Solifugae) use adhesive organs on their pedipalps for prey capture. J Ethol 29(1):177–180

Wolff J, Huber S, Gorb S (2015a) How to stay on mummy's back: morphological and functional changes of the pretarsus in arachnid postembryonic stages. Arthropod Struct Dev 44:301–312

Wolff JO, Seiter M, Gorb SN (2015b) Functional anatomy of the pretarsus in whip spiders (Arachnida, Amblypygi). Arthropod Struct Dev 44(6):524–540

Chapter 8
Adhesive Secretions

Abstract The compositions and biological functions of sticky secretions are diverse. We here give an overview about glues in arachnids, distinguishing between viscid glue, which remains viscous and often allows reversible and multiple attachment, and solidifying glue, which creates a durable bonding. Viscid glue is often utilized to capture prey, for example in the capture threads of orb web and cob web spiders, and in the pedipalps of harvestmen. It may also be used to coat eggs to make them tacky for substrate attachment and defence against egg predators. Solidifying glue is represented by some coatings of silk fibres, amorphous mating plugs, egg casings and brood sacs, mouthpart attachment of ticks and some mites, and secretions serving the attachment of soil particles for camouflage, as present in some harvestmen and ticks.

8.1 Introduction

Glue, as we know it from daily experience, is a liquid substance that hardens after application, or a viscous mass on a tape that makes it tacky, a so-called pressure sensitive adhesive (PSA). Similar types of glues are widespread in nature. Also in arachnids we find various kinds of adhesive secretions, fulfilling diverse biological functions.

Glue (an adhesive) is a substance that wets a surface and builds chemical or physical bonds with it, resulting in strong adhesive forces (Kinloch 1987). Efficient glue shows a good wetting behavior on the target surface, strong adhesion (i.e. by building hydrogen or covalent bonds with the substrate surface) and a similarly strong cohesion (inner strength). The initial cohesion should be preferably much lower than the adhesion to the target surface to provide a good wetting behavior, but should then quickly increase to a value at least as high as the adhesion to provide a strong bonding. In glues used for permanent adhesion, as for example in the case of the substrate attachment of sessile organisms, high adhesive strength is usually

---

Electronic supplementary material The online version of this chapter (doi:10.1007/978-3-319-45713-0_8) contains supplementary material, which is available to authorized users Videos can also be accessed at http://www.springerimages.com/videos/978-3-319-45712-3.

J.O. Wolff, S.N. Gorb, *Attachment Structures and Adhesive Secretions in Arachnids*, Biologically-Inspired Systems 7,
DOI 10.1007/978-3-319-45713-0_8

achieved by hardening after application. Hardening may be due to polymerization or the evaporation of a solvent (in biological adhesives usually water) due to the air exposition after secretion (Nachtigall 1974). If the target surface is rough, the bonding strength is additionally increased after hardening due to mechanical interlocking. Permanent biological glues are primarily based on self-assembling proteins or peptides that may cross-link and build a large polymer complex after application and polymerization (Voigt and Gorb 2010; Naldrett 1993; Dickinson et al. 2009; Silverman and Roberto 2007; Hennebert et al. 2008).

The disadvantage of permanent glue is that the formation and break up usually take some time. Hence, they are barely applicable for locomotion, prey capture and defence, which demand a quick action (but see polymeric glue of Onychophorans (Röper 1977)). For these purposes, often viscous or visco-elastic substances are used that may create strong bonding over a short period of time without hardening (Betz and Kölsch 2004). In fluids, cohesion is enhanced by capillary and/or viscous forces. A capillary effect is most efficient in thin films between two surfaces (e.g. an attachment organ and a substrate) (Betz and Kölsch 2004; Dirks and Federle 2011). In thicker films, high capillarity demands a high surface tension, which, however, reduces the spreading behavior on micro rough surfaces. A high viscosity leads to a slow flow, which reduces the ability to separate two conglutinated surfaces, when pulled apart quickly (Betz and Kölsch 2004). Visco-elastic adhesives show a shear thickening behavior, which temporarily highly enhances the cohesion under quick load (Dirks et al. 2009; Sahni et al. 2010; Wolff et al. 2014). In the case of the sundew, a carnivorous plant, this phenomenon is based on a hygroscopic polysaccharide network that is highly compliant under pressure, but stretches to long tough nano-fibres under tension (Huang et al. 2015). Adhesive secretions are based on a large variety of substances, such as mucopolysaccharides (Rost and Schauer 1977), glycoproteins (Vollrath and Tillinghast 1991), emulsions of lipids in water or protein solutions (Voetsch et al. 2002; Betz and Kölsch 2004; Dirks et al. 2009), and waxes or resins (Hermann and Blum 1981). They may contain surfactants that highly reduce the surface tension (Voetsch et al. 2002), which may permit spreading on surfaces with complex microstructures that usually act highly repellent on lots of liquids and are present in both arthropods (Helbig et al. 2011) and plants (Neinhuis and Barthlott 1997).

In this chapter, we review known adhesive secretions in arachnids, however, we do not discuss fluid secretions that support adhesive structures here (refer to the previous chapters and Chap. 10 for those). Some solidifying secretions may be shaped into specific structures, like fibres (silk threads) or mushroom-like structures. In such secretory products, the shape of their contact with the substrate has a high effect on their adhesion, and we discuss these issues separately in the Chaps. 4 and 5.

8.2 Viscid Glue

8.2.1 Viscid Silk

The major lineage of orb web spiders and relatives, comprising some of the largest spider families, the Araneidae and Theridiidae, uses glue coated threads to capture insect prey. The viscid glue is a secretion of the aggregate glands containing hygroscopic glycoproteins (Vollrath and Tillinghast 1991) and organic and inorganic salts (Vollrath et al. 1990; Sahni et al. 2014b). The glycoproteins bind atmospheric water, forming viscoelastic mucilage (Sahni et al. 2010, 2011; Edmonds and Vollrath 1992) (Fig. 8.1c). The salts enhance the solvation of the glycoprotein, and thus permit good ability to wet substrates, even at low humidity (Sahni et al. 2014b). The aggregate gland secretion coats a simultaneously spun fibre of flagelliform silk. The coating is achieved by a very close location of the aggregate and flagelliform spigots (Eberhard 2010) (Fig. 8.1b). The initially cylindrical coat quickly breaks up into regularly arranged droplets (so-called beads-on-a-string (BOAS) morphology), due to the capillarity driven Raleigh instability (Torres et al. 2014) (Fig. 8.1a). This causes the internal flagelliform fibres to coil within the droplets. Under load the viscid glue coat is stretched and the fibres de-coil, which permits an extreme extensibility of the capture thread (Blackledge and Hayashi 2006; Elettro et al. 2016). By the high extensibility of both the thread and the viscid glue a flying prey can effectively be stopped without a failure of the glue bond. The stickiness of the glue is assumed to be constrained by the strength of the flagelliform fibres, because an excessive stickiness would not allow the spider to take the prey item without destroying its web (Agnarsson and Blackledge 2009). It is further constrained by the inner strength of the glue, which is always significantly higher than the adhesion, presumably to avoid glue loss and keeping the functionality of the capture thread (Opell et al. 2011). Nonetheless, the viscid glue is so effective that in some lineages the orb web structure has been highly reduced. In bolas spiders (Araneidae: *Mastophora*) only a single thread is spun carrying a single glue droplet, which is actively thrown onto moths flying by (Eberhard 1980). Theridiidae use aggregate glue in so-called gumfoot traps (Fig. 8.1d) and also throw threads with a viscid glue coating directly onto the prey (Coddington 1989; Hajer and Hrubá 2007). These spiders also use the glue in defence against aggressors and predators (Vetter 1980). In both gumfoot and attack/defence threads, the viscid glue comes from a separate aggregate gland (Coddington 1989), and may thus differ in its drying and adhesion properties. However, this has not been studied to date. Similar gumfoot traps and sticky silk attacks have also been reported from Pholcidae, although these do not have aggregate glands (Japyassú and Macagnan 2004) (Fig. 8.1e–g). Hence, it is a convergent development, presumably on the basis of a derived piriform gland (Japyassú and Macagnan 2004).

A rather unknown analogue of viscid capture threads may occur in trombidiform mites belonging to the family Cunaxidae. These spin silk threads covered with a viscoelastic glue coat forming BOAS structures (Fig. 8.7e, f). The threads are woven

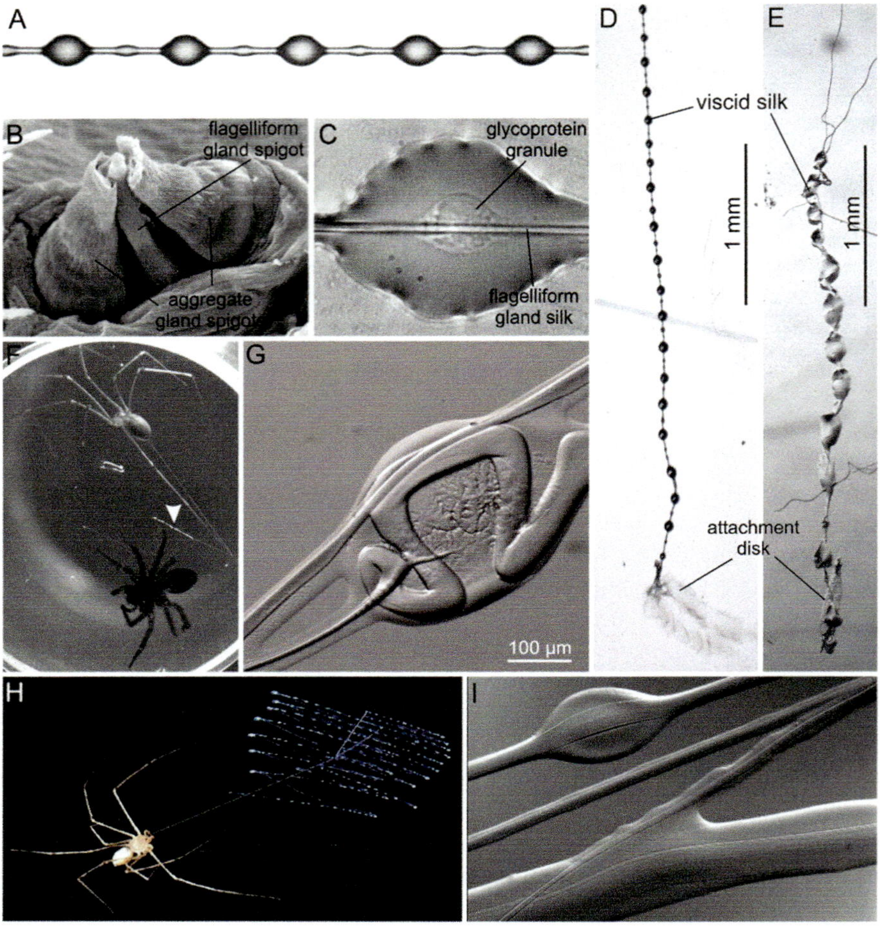

Fig. 8.1 Viscid silk of araneomorph spiders. (**a**–**c**) The capture threads of araneid orb web spiders (here: *Argiope trifasciata*) consist of the extremely extensible flagelliform gland silk, coated with the viscoelastic glue of the aggregate glands, which forms BOAS structures (after Sahni et al. 2014a); (**b**) the coating is achieved by the close proximity of two aggregate gland spigots enclosing the flagelliform gland spigot (this spigot group is also called the *triad*); (**c**) a DIC image of a glue droplet shows the internal glycoprotein granule and the viscid coat. (**d**) Cobweb spiders (Theridiidae, here: *Parasteatoda tepidariorum*) build so-called gumfoot-traps to capture walking prey, like ants; these are pre-stressed glue coated threads anchored to the ground via a special type of attachment disc that breaks when the prey walks against the trap. (**e**) Similar gumfoot traps have recently been found in the non-related daddy-longlegs spiders (Pholcidae, here: *Pholcus phalangioides*); the viscid glue coat presumably comes from a modified piriform gland. (**f**) These spiders also throw glue coated threads directly onto prey and predators; here a *P. phalangioides* is attacking an *Amaurobius fenestralis* by pulling sticky threads from its spinnerets with its hind legs, trying to place it on the opponent (viewed from *below*; *arrowhead* indicates where the sticky thread sticks to the underlying glass slide). (**g**) Detail of the deposited glue on the glass slide after the attack; it is composed of an internal fibre and a glue coat. (**h**, **i**) The spitting spider *Scytodes thoracica* (Scytodidae) ejects a fibre-glue composite from its chelicerae onto the prey in a zigzag-pattern (after Suter and Stratton 2009)

into three-dimensional snares, in which the mite rests ambushing (Alberti and Ehrnsberger 1977). However, although it was observed that collembolan prey gets entangled in the web and is then overwhelmed by the mite, the stickiness of the BOAS structures remains an assumption, since the authors did not test the adhesiveness of the threads (G. Alberti, pers. comm.). As discussed below for spider piriform gland silk, a coat of solidifying glue can similarly develop BOAS structures, when exhibiting viscoelastic properties in its initial fluid phase and when it does not contact a substrate. So, the stickiness of the cunaxid silk remains to be tested in future studies.

8.2.2 *Ejection of Viscid Glue onto Prey*

Spitting spiders (Scytodidae) eject a contractile gluey substance directly on the prey (McAlister 1960). The spit is a viscous fluid containing proteinaceous fibres and fibrous glycoproteins (Fig. 8.1i), originating from the highly enlarged venom glands (Kovoor and Zylberberg 1972). It is sprayed in a zig-zag pattern onto the prey, which is a result of an oscillating movement of the chelicerae produced by the hydrostatic pressure of the ejected fluid (Suter and Stratton 2009) (Fig. 8.1h). Shortly after ejection (after 300 ms) the secretion highly contracts (shortening of about 60 %), which strongly ties the prey down (Suter and Stratton 2009). The physical effect causing the adhesion and contraction is not fully understood yet.

8.2.3 *Viscid Prey Capture Glue in Harvestmen*

Specialized glandular hairs on the pedipalps are known from the Dyspnoi and Eupnoi (Gruber 1970; Lopez et al. 1980; Shear 1986; Giribet et al. 2002; Wolff et al. 2016b). They exhibit a rather stiff stalk and a flexible apical section ('plumose part') which is non-slerotized and equipped with numerous hooked microtrichia that hold a droplet of viscid secretion (Fig. 8.2d–g). Both the microtrichia and their apical hooks branch off perpendicularly (Fig. 8.2f), which hampers the fluid flow under tensile load. This efficiently ensures that the glue stays attached. In the higher Dyspnoi (Nemastomatidae, Dicranolasmatidae and Nipponopsalididae) the plumose section is further modified: the microtrichia are radially arranged, which permits the attachment of a spherical droplet (these setae are called clavate or '*Kugelhaare*', (Rimsky-Korsakow 1924; Wachmann 1970)) (Fig. 8.2d–f). Another feature that evolved among Dyspnoi are slit like openings of the secretion delivery channels (Fig. 8.2f), which presumably permits a quicker extrusion of a more viscous fluid (Wolff et al. 2016b). The glandular setae are used to capture small, fast moving prey like springtails or leaf hoppers (Wolff et al. 2014; 2016b) (Fig. 8.2c; Videos 8.1, 8.2, 8.3 and 8.4). The fluid has viscoelastic properties and can produce high adhesive forces both on glass and springtail cuticle (Wolff et al. 2014) (Video 8.6).

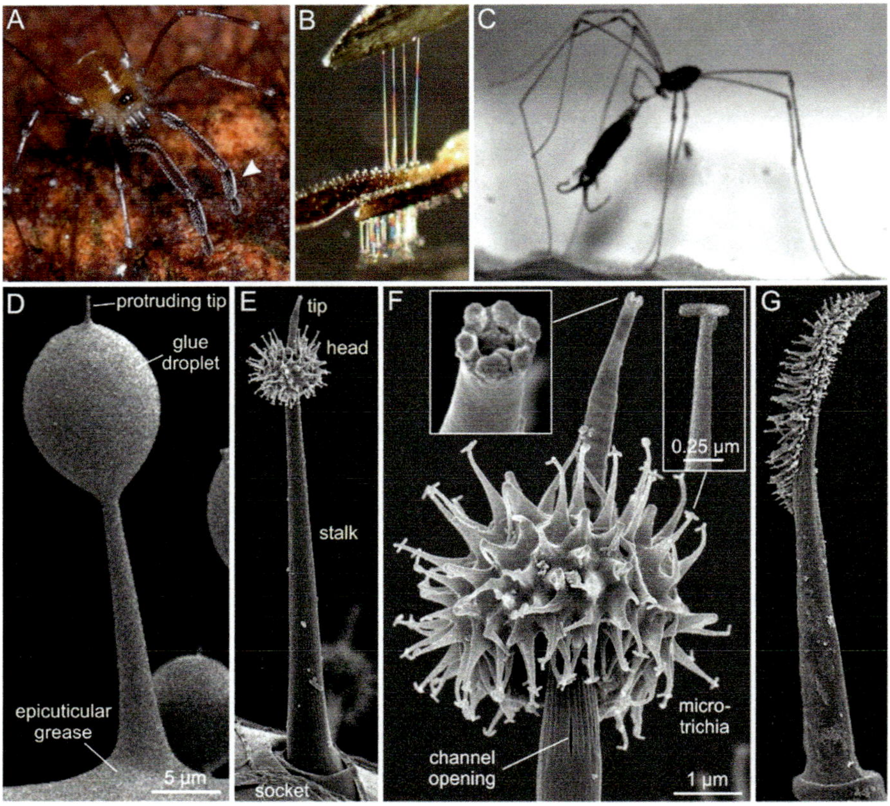

Fig. 8.2 Pedipalpal glandular setae in harvestmen. (**a**) Many harvestmen from the temperate zones bear specialized setae on their pedipalps that carry a glue droplet at their tip, like this juvenile *Mitostoma chrysomelas* (Nemastomatidae). (**b**) The glue is viscoelastic and can be stretched into long fibres (also see Videos 8.5 and 8.6). (**c**) The glue is used to capture agile prey, like in this *M. chrysomelas* capturing an entomobryomorph springtail (also see Videos 8.1 and 8.2). (**d**) Cryo-scanning electron micrograph (Cryo-SEM) of a clavate seta of *M. chrysomelas*; note the setal tip protruding the glue droplet and the presence of epicuticular fluids that are washed off in (**e**) and may act as an anti-sticky coating. (**e, f**) Same after washing with acetone, followed by critical point drying, revealing the ultrastructural details of this specialized type of seta; (**f**) detail of the microtrichious head and tip. The adhesive fluid emerges from slit-like openings underneath the head. Insets demonstrate details of the tip (apical view), showing pores of a chemosensor; and the branching tip of a microtrichium. (**g**) Plumose seta of *Thrasychirus gulosus* (Neopilionidae); note an anisotropic distribution of microtrichia in the upper half of the seta. After Wolff et al. 2016b

Viscid glue for prey capture has independently evolved in the other large suborder of harvestmen, the Laniatores. Nymphs of Cosmetidae have modified pedipalps which lack any spines and exhibit a bulbous pretarsus that lacks the claw (Wolff et al. 2016b). Ultrastructural examination of the pedipalp in a nymph of *Cynortellana quadrimaculata* (Cosmetidae) revealed that the distal patella, the tibia, tarsus and the modified pretarsus are covered with tube-like gland openings with remains of some amorphous secretion. Some recent observations indicated that the pedipalps in these nymphs are covered with a sticky mucous, which is presumably used for prey capture (G. Machado, pers. comm.). Similar glue was found in nymphs of the related Gonyleptidae, especially in the sub-family Cranainae (Fig. 8.3), although their pedipalps are not modified (Wolff et al. 2016a). Their glue exhibits viscoelastic properties and a high water content (Wolff et al. 2016a).

8.2.4 Glandular Setae in Mites

The predatory Epicriidae (Mesostigmata) have glandular setae in their front legs carrying droplets of a viscous secretion used for prey capture (Alberti 2010). Comparable to the glandular setae of harvestmen, which have a similar function (see above), these setae have internal secretion channels and bear microtrichia at the tip, where the secretion droplet is attached (Alberti 2010).

8.2.5 Egg Casings and Brood Sacs

In many harvestmen, the eggs are covered by mucus (viscous hydrogel based on glycoproteins and polysaccharides), during or after their deposition (Juberthie 1964; Witaliński and Żuwała 1981; Requena et al. 2009). The mucus facilitates attachment to substrates, like leaves or stones (S. García-Hernández, pers. comm.). In some species of Gonyleptidae, a large patch of mucus is produced (Fig. 8.8b), which, besides the attachment function, also serves as an effective barrier against egg predators (Requena et al. 2009). These secretions are obviously highly

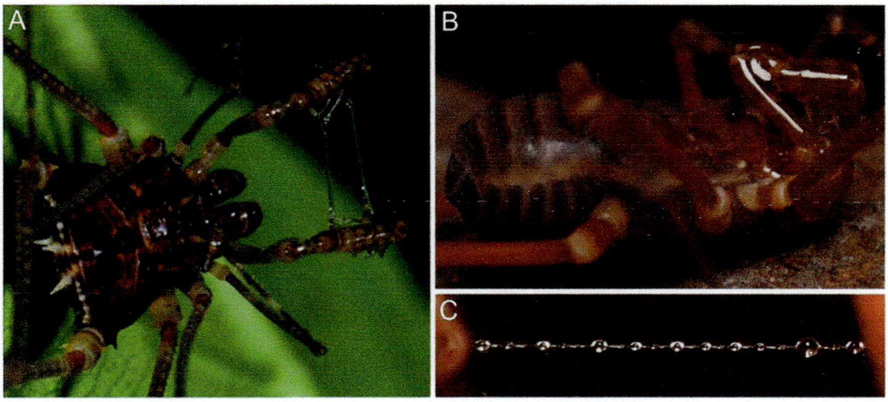

Fig. 8.3 Pedipalpal glue in Laniatores nymphs. (**a**) Nymph of an undetermined species of Cranainae (Gonyleptidae), sitting in an ambushing position. Note the viscid glue on the pedipalps (Photo: A. Anker, with kind permission). (**b**) Nymph of *Phalangodus briareos*; note the much higher reflectance of the pedipalps indicating the glue cover (Photo: J. Uribe, with kind permission). (**c**) The glue on the pedipalps can be pulled into long threads that form BOAS structures over time (Photo: J. Uribe, with kind permission) (Figure adapted from Wolff et al. 2016a)

hygroscopic and may hold water over a long period of time (Juberthie 1964). Hence, they may, further, serve to keep the eggs in a stable humid environment. Other Gonyleptidae glue their legs separately on the substrate and attach pieces of plants and debris from the surrounding to their viscid coat (Willemart 2001; Stanley 2011).

In pseudoscorpions, the first instars develop in the, so-called, brood-sac, which is a pad of mucus attached at the underside of the opisthosoma and covered by a membrane of dried secretion (Weygoldt 1968). The mother secretes nutrients into the brood sac that are absorbed by the larvae (Weygoldt 1968).

8.3 Solidifying Glue

8.3.1 Silk with a Solidifying Glue Coat

Piriform Gland Silk of Araneomorph Spiders Whereas the prey capture threads of spiders are adapted to stick quickly and reversibly, others have to ensure stable bonding. In the Araneomorphae this is achieved by piriform gland silk. This silk is composed of a tough central fibre and a viscoelastic coat that quickly hardens after application, cementing the thread with a substrate and/or other threads (Wolff et al. 2015) (Fig. 8.4c, d). Both the fibre and the cement are produced by histochemically distinct gland tissues and do not mix within the gland lumen (Kovoor and Zylberberg 1980). The cement consists of numerous aligned nano-fibrils, lipid enclosures and an isotropic layer assembling at the interface between substrates and the air (Wolff et al. 2015).

The composition and physical function of the glue are not fully understood yet, but hydrogen bonds with an adsorbed water film on the substrate may play a substantial role, since adhesion is reduced by a strongly hydrophobic coating of the substrate (Wolff et al. 2015). Piriform gland silk fibres are usually not spun separately but in dense clusters, so called attachment discs (Fig. 8.4a), which exhibit an overall mushroom-like geometry (discussed in Chap. 5). This specific way of applying the glue highly enhances the compliancy and flaw tolerance in the bonding and therefore the adhesive strength of the anchorage (Wolff et al. 2015) (see Chap. 5).

Highly enlarged piriform glands occur in the Gnaphosidae (Fig. 8.4c), which do not use the secretion for attachment discs, but to tie down dangerous prey (e.g. other spiders) (Grimm 1985; and pers. obs.). Female wolf spiders (Lycosidae) also use the piriform gland silk to attach the egg sac to their spinnerets as a part of the maternal care (Townley and Tillinghast 2003).

Cylindrical Gland Silk of Araneomorph Spiders In egg sacs of spiders, cylindrical gland silk forms a dense protective cover (Foelix 2011). These threads are composed of numerous fibres within a matrix that appears electron dense in the transmission electron microscope (Stubbs et al. 1992), similar to piriform gland silk or the silk of silkworm cocoons. The matrix may still be fluid during extrusion and conglutinate different threads, resulting in a tough meshwork. However, this has not yet been studied in detail.

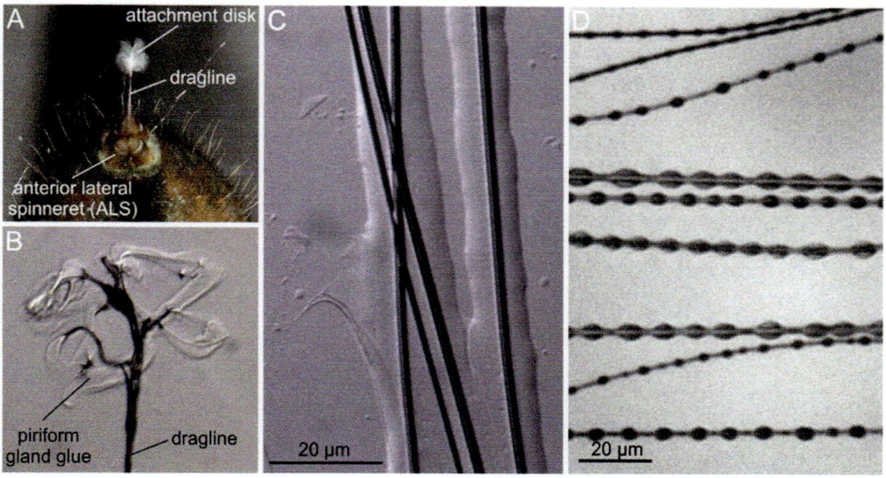

Fig. 8.4 Piriform gland silk of araneomorph spiders. (**a**) Both web-building and free hunting spiders (here: *Cupiennius salei* (Ctenidae)) anchor their silk threads to substrates with sticky silk from the piriform glands, applied in a disc-like structure. (**b**) The proportion of glue differs between spider families; in the Pholcidae (here: *Pholcus phalangioides*) it is comparably high. (**c**) In Gnaphosidae (here: *Scotophaeus* sp.), the piriform glands and spigots are highly enlarged, resulting in larger silk trails, making the usual composition of proteinacous fibre and glue coat well visible in the light microscope. (**d**) When the piriform gland silk is suspended to the air for some time before its application to the substrate (e.g., when the spider, here the crab spider *Thomisus onustus*, is interrupted in spinning) the glue coat forms BOAS structures, indicating its viscoelastic properties prior solidification; the thin threads in the top are single piriform gland fibres of normal size, thicker threads with larger droplets are bundles of conglutinated fibres from several spigots

Silk of Liphistiid and Mygalomorph Spiders Mygalomorphs and the primitive Mesothelae lack the high functional diversity of silk glands. However, some lineages, especially the Atypidae, Dipluridae, Hexathelidae and some Theraphosidae may construct large webs or silken tubes. This implicates that the threads are somehow conglutinated and fastened at the ground. Histochemical studies on mygalomorph silk glands show a separate staining of two gland tissues similar to the piriform glands of araneomorphs (Glatz 1973; Palmer et al. 1982). This indicates that some mygalomorph silk fibres may also exhibit a cement coat. However, studies on the adhesion of mygalomorph silk are not available. We studied the silk of *Liphistius birmanicus* (Fig. 8.5f–h) and *L. ornatus* (Fig. 8.5i) (Liphistiidae), *Atypus piceus* (Fig. 8.6a–c) (Atypidae), *Cteniza* sp. of Sardinia (Fig. 8.5b–e) (Ctenizidae), *Idioctis* sp. of the Philippines (Barychelidae), and *Macrothele* sp. from Turkey (Fig. 8.6d, e) (Hexathelidae), with means of light and scanning electron microscopy.

In these spiders, the silk is used in the burrow linings and trapdoor structures. In both linings and trapdoors, the threads are conglutinated and soil particles are included in the web, partly attached by entanglement, and partly by agglutination of the silk. The trapdoors of Liphistiidae, Ctenizidae and Barychelidae are lid-like structures that perfectly fit onto the opening of the burrow and are covered by debris and plant material of the surrounding, which makes them nearly invisible, when closed (Fig. 8.5a–c).

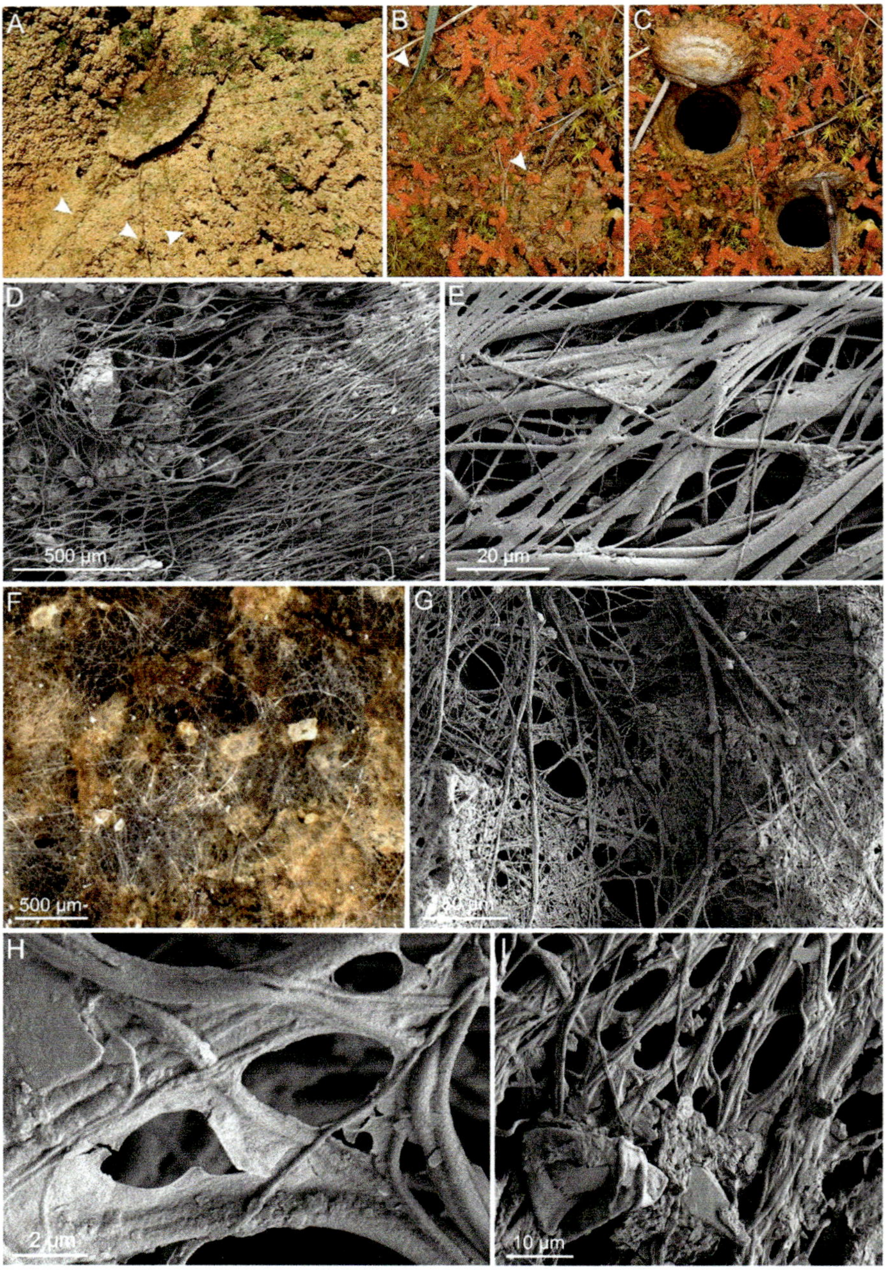

Fig. 8.5 Silk of trapdoor spiders. The most ancient spider lineages build burrows lined with silk and closed with a lid (trapdoor). (**a**) The basal Liphistiidae (here: *Liphistius* cf. *dangrek*) build radially-arranged stumbling lines (*arrowheads*) connected to the burrow; when a prey walks over these the spider gets alarmed and flips out of the burrow to grab it. (**b**, **c**) The trapdoors of the Sardinian *Cteniza* sp. (Ctenizidae) are well camouflaged and hard to spot (*arrowheads* in (**b**)); the burrows and trapdoors are hold in shape by the lining of gluey silk (see (**c**)). (**d**, **e**) SEM reveals that the silk threads of these trapdoors are glue coated, attaching soil particles and creating a tough network. (**f**–**h**) In the liphistiid trapdoors (here: *L. birmanicus*) thin glue coated, thick non-coated threads, and soil particles are interwoven. (**i**) Same accounts for the trapdoors of *L. ornatus*

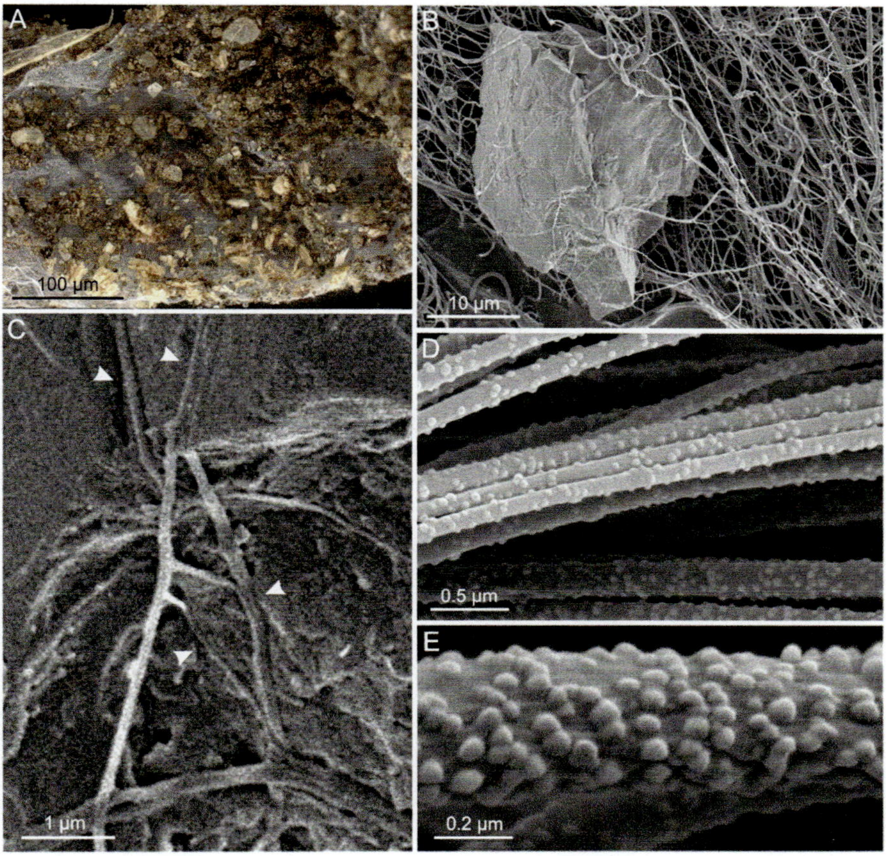

Fig. 8.6 Silk of Atypidae and Hexathelidae. (**a–c**) The Atypidae (here: *Atypus piceus*) build long extensions of the burrow linings above the ground, camouflaged with soil particles; the silk does not exhibit a discernible glue coat, yet it sticks to the soil particles, like sand grains (see *arrowheads* in **c**). (**d**, **e**) The Hexathelidae (here: *Macrothele* sp. from Turkey) build extensive sheet webs, their silk is thin, dry and exhibits a nubby surface

The lid is very stable, since it must protect the spider from predators. It is attached to the ground and lining with a joint made of silk threads. Both features indicate the extensive use of a solidifying glue. In *Liphistius* spp., the webs exhibit thin, clearly glue-coated, and thick, non-coated threads (Fig. 8.5g), which is a configuration quite comparable to that of silken products in araneomorphs. However, these spiders do not spin attachment discs and draglines, but use these two types of silk to build trapdoors, to line their burrows and to produce signal lines extending from the burrow and being attached to the substrate (Fig. 8.5a). The trapdoors of *Cteniza* and *Idioctis* exhibit a very similar composition of fibres and glue (Fig. 8.5d).

In contrast, the fibres of *Atypus* and *Macrothele* are only loosely connected and might act more like a dry adhesive. They are relatively thin and do not exhibit a discernible glue coat (Fig. 8.6). In case of *Atypus,* this is presumably attributed to the specific function of the lining. In these spiders, the lining extends above the ground, forming a tube. When a prey walks over this tube, the spider bites from the inside through the web and tears it open, to pull the prey into the burrow (Foelix 2011). If the threads in the tube web would be as strongly agglutinated as in the linings of trapdoor spiders, the spiders would have difficulties to perform this specific way of prey capture. Yet soil particles stick to the silk (Fig. 8.6c), but it is unclear by which mechanism. The silk threads of *Macrothele* exhibit a nubby surface (Fig. 8.6d, e) – these nubs may represent contact points, supporting a dry adhesive function as in cribellar silk (see Chap. 4) or act as a Lotus-effect-like surface making the web strongly hydrophobic. These spiders do not build burrows, but rather extensive sheet webs in hollow trees. It remains unclear, whether another type of silk, that is glue-coated, is used to attach the threads to the substrate, because we only studied a small piece of the web here.

Tarsal 'Silk' In climbing tarantulas (Theraphosidae) thread-like traces originating from the feet have been reported (Gorb et al. 2006; Rind et al. 2011; Peattie et al. 2011). Those were first interpreted as primitive silk threads which should enhance the attachment ability (Gorb et al. 2006). This was rebutted by the finding that the secretions emerge from chemosensory setae and remain fluid (Foelix et al. 2012, 2013). The thread-like appearance comes from the dragging movement during locomotion or holding onto vertical surfaces. Tarsal chemosensors with a distinct distal pore are ubiquitous in spiders. Meanwhile, we made similar observations of thread-like tarsal secretions with cob web spiders (Theridiidae) and sac spiders (Clubionidae), showing that this is not restricted to mygalomorphs. We assume that the foot secretion does not serve adhesion, but rather functions as a solvent for chemical substances, which can be sensed by gustatory receptors. Nonetheless, this observation shows that secretory chemosensors may have played a role in the evolution of spigots and silk.

Silk of Other Arachnids Some mites and pseudoscorpions are able to produce fine silk threads, too. Spider mites (Trombidiformes: Tetranychidae) bear spigots on their pedipalps. The silk is used to build silken covers as a protection against predators, rain and wind, and to exclude interspecific competitors (Gerson 1979). In the presence of predatory mites, eggs are laid suspended in the webbing, where they are less imperiled by predation (Lemos et al. 2010). As these mites live in large colonies, the silk threads of individuals are connected to form large cover sheets. The amino acid composition of the spider mite silk is comparable to that of pseudoscorpions (Hazan et al. 1975). With a diameter of 50–100 nm, its fibres are much thinner than the ones of other spinning arthropods (Clotuche et al. 2012). Hence, the fibres might significantly adhere to substrates or between each other by van der Waals forces or the interlocking with minute surface asperities. However, it was also observed that spinning mites shortly remain still with the pedipalps pressed onto the substrates, in order to attach the silk (Kanazawa et al. 2011; Clotuche et al. 2012). This may indicate that the silk is still fluid, when emerging. This is also plausible from the rheological point of view, because the flow through the extremely narrow duct would be impossible, if the secretion does not keep a low viscosity

(Elettro et al. 2015). The threads of spider mites seem to stay adhesive for up to 1 h, which may indicate that some slowly drying fluid coating is present on the threads (Kanazawa et al. 2011). The mites use this property to remove dirt particle from their nest (Kanazawa et al. 2011).

A highly sticky silk is produced by snout mites (Trombidiformes: Bdellidae), which use it to immobilize prey and to build nests (Alberti 1973). The silk is adhesive after emergence, and dries over time, thereby losing its stickiness (Alberti and Ehrnsberger 1977). It is clearly a bi-phasic secretion with the central component forming a fibre and the coat of solidifying glue (Figs. 8.7c, d). For prey capture, larger amounts of the secretion are ejected onto the prey, from which a thread is pulled and attached to the ground (Alberti and Ehrnsberger 1977). By this means, highly mobile prey items, like springtails or other mites, are tethered to the substrate and overwhelmed (Alberti 1973) (Fig. 8.7a). The glue exhibits very good wetting properties, spreading over the complex Lotus-leaf-like cuticular surfaces of springtails (Fig. 8.7b) and even oribatid soil mites (Fig. 8.7a), which have a strongly repellent hydrocarbon coating (Raspotnig and Leis 2009). The Cunaxidae, which are closely related to the Bdellidae also exhibit a biphasic silk, which is used to build prey capture webs (Alberti and Ehrnsberger 1977) (Fig. 8.7e, f, and see discussion in Sect. 8.2).

In pseudoscorpions, the silk glands are located in the prosoma and open into spinnerets onto the chelicerae (Weygoldt 1969). The silk is used to produce protective retreats while breeding, molting or hibernating (Kovoor 1987). Males of *Serianus carolinensis* (Olpiidae) also use silk to mark deposited spermatophores for females (Weygoldt 1966). It is unknown how the silk is attached to substrates or conglutinated, but we assume that the mechanism is similar to the delicate fibres of spider mites.

8.3.2 Egg Cases

In amblypygids and thelyphonids, the eggs are not deposited on substrates but carried at the underside of the opisthosoma (Fig. 8.8a). They emerge embedded into a viscous filamentous fluid that hardens and gets tanned after being attached at the opisthosoma, especially in the whip-spiders (Weygoldt 2000). This, so-called, egg sac has been hypothesized to present a link to silk evolution in spiders (Pocock 1895), which, however, was not supported by more recent authors, because the amblypygid egg sac glands are mesodermal, whereas the spider silk glands are ectodermal (Decae 1984; Shultz 1987). The paired egg sacs of Thelyphonida (Fig. 8.8a) exhibit only a weak dried and hardened protective shell and contain fewer fibres, and might thus be more primitive than those of amblypygids (Kästner 1941; Weygoldt and Huber 2013).

In some harvestmen, the mucous egg cover (see above) does not stay viscous, but dries, presumably due to a different composition. This strongly cements the eggs to

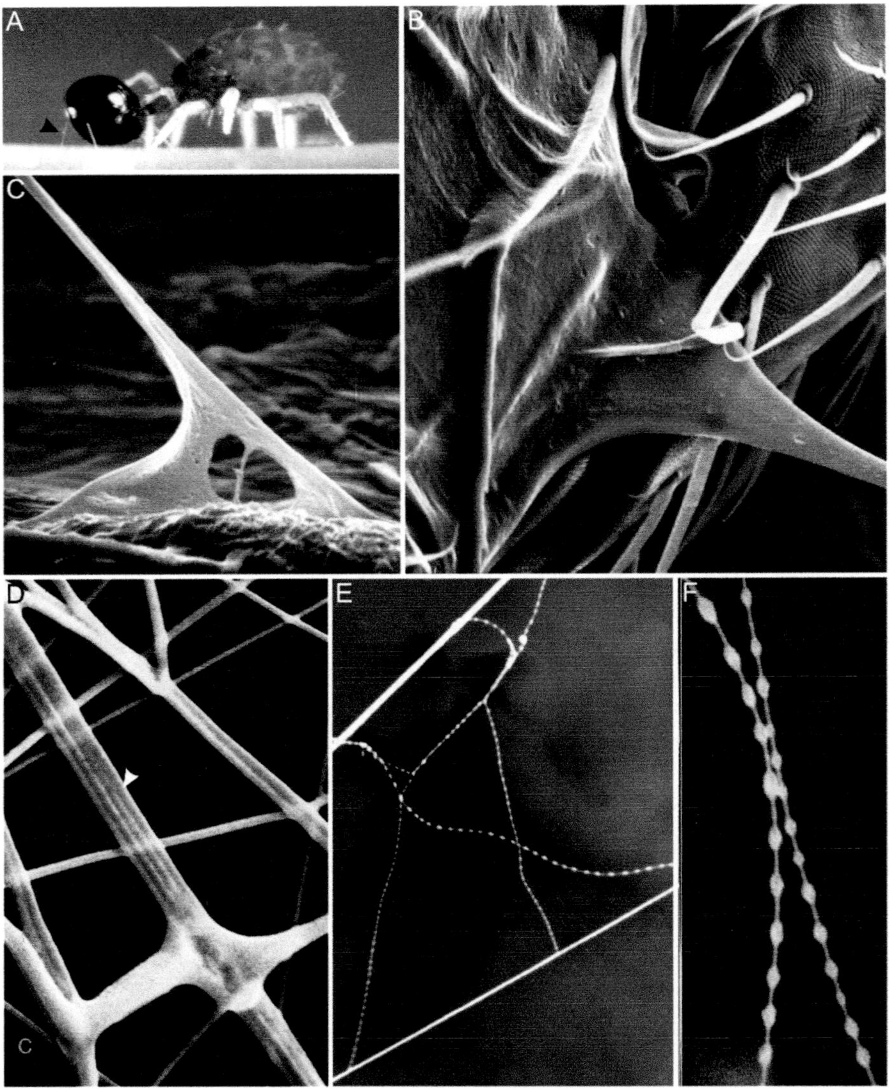

Fig. 8.7 Sticky silk of trombidiform mites. (**a**) Snout mites (Bdellidae) spin sticky silk, which is used to tether prey to the ground (here a *Cyta latirostris* capturing an oribatid mite with the arrowhead pointing to a tethering thread, after Alberti 1973, with kind permission by G. Alberti). (**b**) The glue coat highly spreads on the complex Lotus-effect-like cuticle of springtails (here the secretion of *Biscirus silvaticus*). (**c**) This SEM micrograph shows the ground substrate attachment of the same tethering thread. (**d**) Detail of retreat web of *Biscirus silvaticus*, arrowhead points to a bundle of parallel fibres clearly distinguishable from the glue coat. (**e**, **f**) The Cunaxidae (here: *Cunaxa setirostris*) build capture webs assisting in prey retention and including delicate fibrils with a beads-on-a-string structure, indicating the presence of a viscoelastic glue coat (Images (**b**–**f**) after Alberti and Ehrnsberger (1977), with kind permission by G. Alberti)

Fig. 8.8 Adhesive egg casings in arachnids. (**a**) Whip-scorpions (here: *Typopeltis magnificus*) cover their eggs with a sticky mucus and attach them to the underside of their abdomen (Photo: S. Huber, with kind permission). (**b**) Some harvestmen (here: *Iporangaia pustulosa* (Gonyleptidae)) cover their egg clutches by mucus, in order to attach them to the substrate and protect them from egg predation; note that the mucus coat remains stable and sticky over a long period of time, since the eggs already show a late developmental stage (Photo: G. Requena, with kind permission). (**c**) The eggs of *Phalangodus briareos* (Gonyleptidae) are glued onto a wet rock surface with means of a solidifying secretion (arrowheads point to plaques of the adhesive) (Photo: J. Uribe, with kind permission). (**d**) Males of *Leytpodoctis oviger* (Podoctidae), carry eggs glued to the right femur of the leg IV (Photo: J. Martens & J. Trautner, with kind permission) (Images **b–d** adapted from Wolff et al. 2016a)

their substrates, i.e. in the gonyleptid *Phalangodus briareos*, which glues its eggs onto humid rock surfaces in caves (Wolff et al. 2016a) (Fig. 8.8c), or in *Leytpodoctis oviger* (Podoctidae), in which the eggs are glued onto the male's leg by the female (Martens 1993) (Fig. 8.8d).

8.3.3 Amorphous Mating Plugs

In many spiders, males seal the female genital opening after copulation, in order to avoid sperm concurrence and secure their paternity (Uhl et al. 2010). One mechanism is mechanical plugging (see Chap. 2), another the secretion of an amorphous

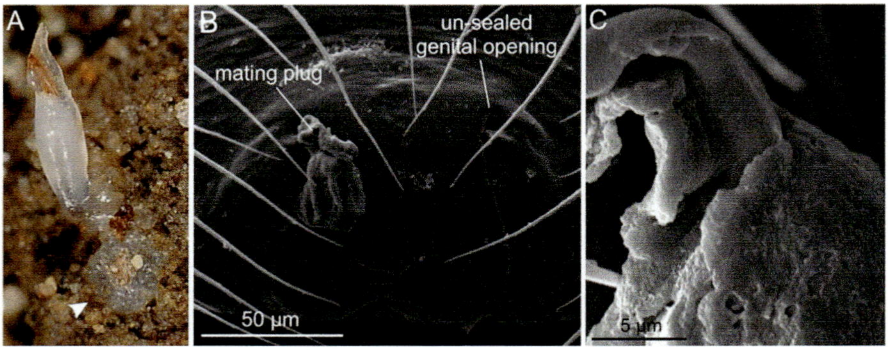

Fig. 8.9 Spermatophores and mating plugs. (**a**) Various arachnids, i.e. scorpions (here *Chaerilus* sp.) use spermatophores for indirect sperm transfer; these are glued onto the substrate (*arrowhead*) via a solidifying secretion applied in a mushroom-like structure (see Chap. 5 for more details). (**b**, **c**) In some arachnids, such as many spiders (here: *Oedothorax retusus* (Linyphiidae)), males seal the female genital opening with a solidifying secretion after copulation (in (**c**) detail of the secretion forming the mating plug) (Micrographs by M. Witthuhn and K. Kunz, with kind permission)

substance that hardens, forming a robust plug that is very hard to remove (Lopez 1987) (Fig. 8.9b, c). The secretion may come from a gland in the pedipalpal bulb (Uhl et al. 2014), or, in some cases, from glands of the female genital tract (Aisenberg and Barrantes 2011). In *Leucauge argyra* (Tetragnathidae), the female secretes the fluid during copulation, which cements the embolus of the copulating male in the genital opening (Aisenberg and Barrantes 2011). The composition and material properties of such plug secretions in spiders are not known yet. Amorphous mating plugs have also been reported from some species of scorpions (Polis and Sissom 1990). In *Euscorpius italicus* (Euscorpiidae), the plug consist of dried sperm (Althaus et al. 2010).

8.3.4 Soil Crypsis

Several arachnids cover their body with a solidifying secretion, in which soil particles are encrusted. This is regarded to be a defensive mechanism, serving camouflage and armoring (Schaider and Raspotnig 2009). Some harvestmen can blend in their environment, where they are nearly not discernible, because of a cover with soil particles. This is particularly well known for the Trogulidae (Schwangart 1907) (Fig. 8.10a, b), but also for Dicranolasmatidae (Gruber 1993) (Figs. 8.10g, h), some Nemastomatidae (*Ortholasma*, *Dendrolasma*, *Vestiferum*) (Schönhofer 2013), some Sclerosomatidae (*Sclerosoma* spp.) (Gnaspini and Hara 2007), some Triaenonychidae (Gnaspini and Hara 2007), some Phalangodidae (Roewer 1923), some Podoctidae (Martens 1993) (Fig. 8.8d), and some Gonyleptidae (Hernandariinae, Ampycinae, some Pachylinae) (DaSilva and Pinto-da-Rocha 2010). The particles are attached by a hardening (presumably tanning) secretion. In *Trogulus* spp., the secretion is produced and stored in bubble-like cuticular protuberances (*'Drüsenwärzchen'* = glandular warts) (Schwangart 1907) (Fig. 8.10c, e). We examined a freshly molted individual by means of scanning electron microscopy. There were no discernable pores in the warts nor between them, hence the secretion is presumably released by diffusion trough their thinned cuticular walls. The hardened secretion forms a very robust crust with embedded sand grains and detritus, which can hardly be removed. Short, hook-like curved setae on some warts (Fig. 8.10e) additionally help to catch and retain soil particles. The only clean body parts are the tarsi, eyes, and mouth parts: the latter are remarkably free of dirt. In *Dicranolasma* spp., no glandular warts are present. The secretion assembles as a film on the cuticle and forms clumps at the short body setae (Fig. 8.10i). The secretion seems to contain a fibrous material (Fig. 8.10j). In nymphs, the crust is comparably easy to remove, indicating a lower amount or divergent composition of the secretion. Nonetheless, large clumps of soil material assemble on the dorsum. The composition and properties of these secretions are not known.

Like these harvestmen, some mites secrete a hardening fluid, in order to cover their bodies with soil particles for camouflage. In particular, we found this in the Caeculidae (Trombidiformes) (Fig. 8.11a). The secretion coagulates on the site of extrusion (Fig. 8.11c) and appears dark brown and tanned. On parts of the ventral side and at the arthrodial membranes, no secretion is present. The underlying cuticle exhibits elaborate microstructures, forming hexagonal combs and globular granules (Fig. 8.11d). These parts are totally free of any dirt, which is consistent with the observation that very similar microstructures in springtails work fluid-repellent and self-cleaning (Helbig et al. 2011). In *Caeculus,* there seems to be a dual function of these structures: they may provide good hold for the coagulating secretion, but prevent adhesion of soil particles and fluids at places, where it would disturb sensing or motion. Soil crypsis also occurs in larvae of some Oribatida, like *Belba corynopus* (Damaeidae) (Seniczak and Seniczak 2013) (Fig. 8.11b). In some species, the exuvia (in some cases with soil particles) stays attached for camouflage (Seniczak and Seniczak 2013). In all these cases, nothing is known about the composition and properties of the secretions.

8.3.5 Cementing of Mouthparts in Ticks and Mites

Ticks (Ixodida) secrete a proteinaceous cement cone with carbohydrate and lipid additions, to anchor their mouthparts in the host's skin (Gregson 1960; Kemp et al. 1982; Bishop et al. 2002). The hardening is presumably an enzymatic tanning process (Kemp et al. 1982). Together with denticles on the chelicerae sheath, this permits an extremely robust attachment by interlocking with the skin tissue of the host. Interestingly, it is not entirely known, how the ticks lose their attachment, when detaching from their hosts. It was observed, that the cement cone is left in the skin, and it was assumed that the ticks retract their mouthparts from the anchorage (Gregson 1960). A violently removed tick still has cement on its chelicerae, but is

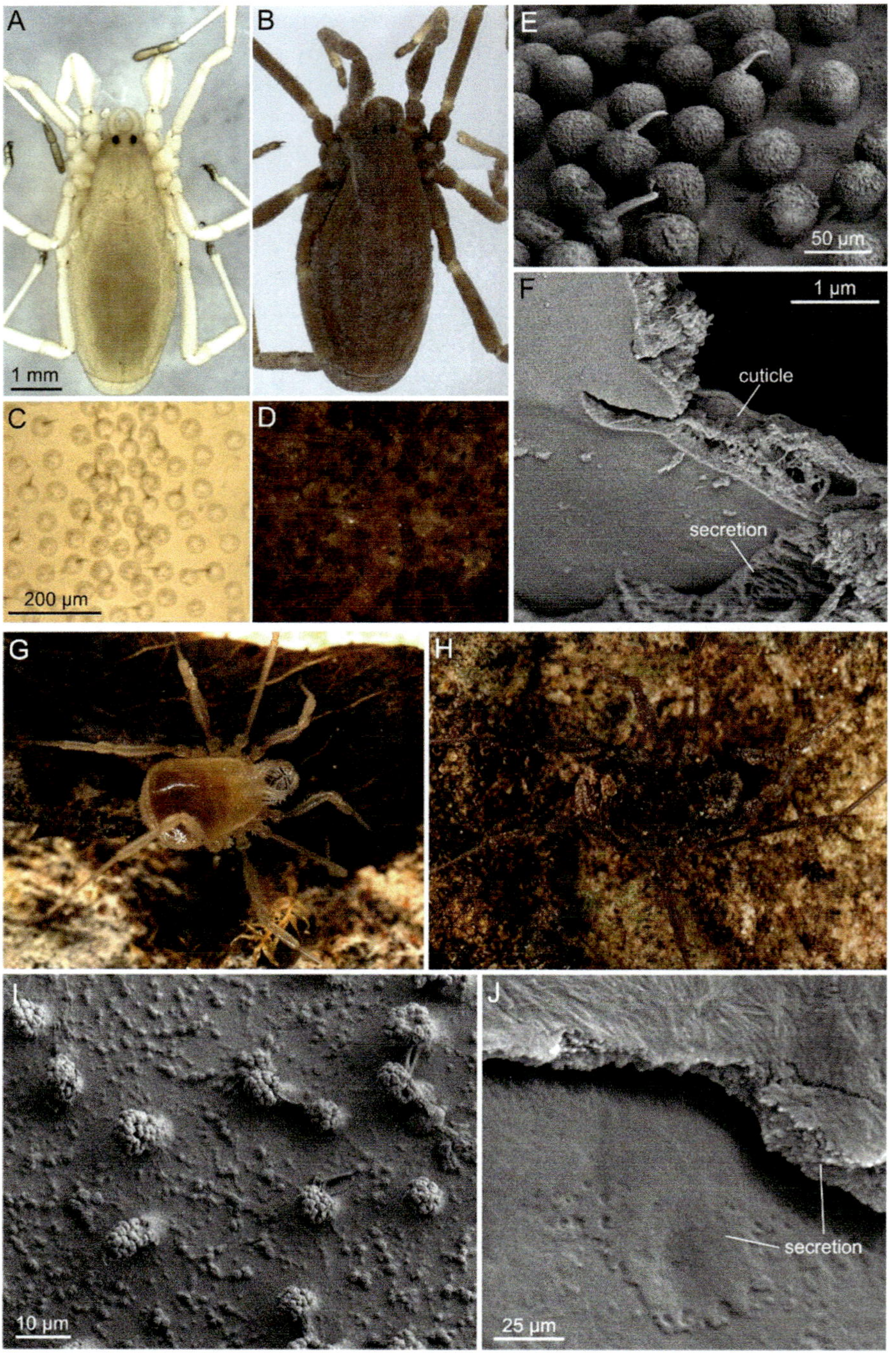

Fig. 8.10 Soil camouflage in Dyspnoi. After molting, some Dyspnoi secrete a solidifying substance, to which soil particles adhere, providing both camouflage and additional armor to the cuticle surface. (**a**–**f**) Soil camouflage in *Trogulus martensi* (Trogulidae); (**a**, **c**, **e**, **f**) freshly molted; and

Fig. 8.11 Soil camouflage in Acariformes. (**a**) Habitus of *Microcaeculus* sp. (Trombidiformes: Caeculidae); note sand grains adhering to the body, and spiny front legs (Photo: A. Grabolle, with kind permission). (**b**) Unidentified Damaeidae (Oribatida) carrying its exuvia with a crust of dried secretion and soil particles (Photo: L. Bendixen, with kind permission). (**c**) SEM of the ventral cuticle of *Caeculus* sp.; note the clumps of dried secretion (*arrowheads*) and the clean ornamented cuticle in between. (**d**) Detail of the micro-ornamentation of the cuticle; the Lotus-effect-like microstructures are presumed both to fasten the dried secretion coat and keep the non-coated sites free from contamination with soil particles

Fig. 8.10 (continued) (**b**, **d**) usual habitus with dried crust of secretion and soil particles; (**c**–**e**) Detail of the dorsal integument with glandular warts; (**c** and **e**) After molting, the warts are still exposed and clearly visible; note the short, curved setae on some warts that assist both the capture and retention of soil particles; (**d**) A secretion emerges and assembles between the warts, catches soil particles and dries, forming a crust that is hardly removable (same magnification and section as in (**c**)); (**f**) The cuticle of a glandular wart, partly cut with a scalpel. The cuticle is smooth, hence the roughness seen in (**e**) is due to already emerged secretion; there are no discernible pores, hence, the secretion may emerge by diffusion through the cuticle. (**g**–**j**) Soil camouflage in *Dicranolasma* sp. (Dicranolasmatidae) (Photos (**g**) and (**h**) by S. Huber, with kind permission); (**g**) freshly molted individual and (**h**) usual habitus with soil cover, photographed in its natural environment; (**i**, **j**) SEM of the dorsal integument of a nymph showing details of the crust of dried secretion; the nubs in (**i**) are clumps of the secretion adhering to setae; glandular warts are absent in this family; (**j**) Fracture of the secretion crust; the underlying smooth cuticle is visible; the secretion seems to contain nano-fibrils able to remove it within a few minutes (Gregson 1960). It cannot be excluded that an enzymatic fluid is used in this process. Adult Argasidae do not secrete the cement anymore and also lack the hooklets on the chelicerae sheath (Kemp et al. 1982).

Such cement cones, anchoring the mouth parts additionally to hook-like structures, have also independently evolved in other mites, such as larvae of *Leptus* spp. (Trombidiformes: Erythraeidae) parasitizing on harvestmen (Åbro 1988).

References

Åbro A (1988) The mode of attachment of mite larvae (*Leptus* spp.) to harvestmen (Opiliones). J Nat Hist 22(1):123–130

Agnarsson I, Blackledge T (2009) Can a spider web be too sticky? Tensile mechanics constrains the evolution of capture spiral stickiness in orb-weaving spiders. J Zool 278(2):134–140

Aisenberg A, Barrantes G (2011) Sexual behavior, cannibalism, and mating plugs as sticky traps in the orb weaver spider *Leucauge argyra* (Tetragnathidae). Naturwissenschaften 98(7):605–613

Alberti G (1973) Ernährungsbiologie und Spinnvermögen der Schnabelmilben (Bdellidae, Trombidiformes). Z Morphol Tiere 76(4):285–338

Alberti G (2010) On predation in Epicriidae (Gamasida, Anactinotrichida) and fine-structural details of their forelegs. Soil Organ 82:179–192

Alberti G, Ehrnsberger R (1977) Rasterelektronenmikroskopische Untersuchungen zum Spinnvermögen der Bdelliden und Cunaxiden (Acari, Prostigmata). Acarologia 19(1):55–61

Althaus S, Jacob A, Graber W, Hofer D, Nentwig W, Kropf C (2010) A double role of sperm in scorpions: The mating plug of *Euscorpius italicus* (Scorpiones: Euscorpiidae) consists of sperm. J Morphol 271(4):383–393

Betz O, Kölsch G (2004) The role of adhesion in prey capture and predator defence in arthropods. Arthropod Struct Dev 33(1):3–30

Bishop R, Lambson B, Wells C, Pandit P, Osaso J, Nkonge C, Morzaria S, Musoke A, Nene V (2002) A cement protein of the tick *Rhipicephalus appendiculatus*, located in the secretory e cell granules of the type III salivary gland acini, induces strong antibody responses in cattle. Int J Parasitol 32(7):833–842

Blackledge TA, Hayashi CY (2006) Silken toolkits: biomechanics of silk fibers spun by the orb web spider *Argiope argentata* (Fabricius 1775). J Exp Biol 209(13):2452–2461

Clotuche G, Mailleux A-C, Deneubourg J-L, Detrain C, Hance T (2012) The silk road of *Tetranychus urticae*: is it a single or a double lane? Exp Appl Acarol 56(4):345–354

Coddington JA (1989) Spinneret silk spigot morphology: evidence for the monophyly of orbweaving spiders, Cyrtophorinae (Araneidae), and the group Theridiidae plus Nesticidae. J Arachnol 17(1):71–95

DaSilva MB, Pinto-da-Rocha R (2010) Systematic review and cladistic analysis of the Hernandariinae (Opiliones: Gonyleptidae). Zoologia (Curitiba) 27(4):577–642

Decae AE (1984) A theory on the origin of spiders and the primitive function of spider silk. J Arachnol 12(1):21–28

Dickinson GH, Vega IE, Wahl KJ, Orihuela B, Beyley V, Rodriguez EN, Everett RK, Bonaventura J, Rittschof D (2009) Barnacle cement: a polymerization model based on evolutionary concepts. J Exp Biol 212(21):3499–3510

Dirks J-H, Federle W (2011) Fluid-based adhesion in insects–principles and challenges. Soft Matter 7(23):11047–11053

Dirks J-H, Clemente CJ, Federle W (2009) Insect tricks: two-phasic foot pad secretion prevents slipping. J R Soc Interface 7(45):587–593

Eberhard WG (1980) The natural history and behavior of the bolas spider *Mastophora dizzydeani* sp. n. (Araneidae). Psyche 87(3–4):143–169

Eberhard WG (2010) Possible functional significance of spigot placement on the spinnerets of spiders. J Arachnol 38(3):407–414

Edmonds DT, Vollrath F (1992) The contribution of atmospheric water vapour to the formation and efficiency of a spider's capture web. Proc R Soc B 248(1322):145–148

Elettro H, Neukirch S, Antkowiak A, Vollrath F (2015) Adhesion of dry and wet electrostatic capture silk of uloborid spider. Sci Nat 102(7–8):41

Elettro H, Neukirch S, Vollrath F, Antkowiak A (2016) In-drop capillary spooling of spider capture thread inspires hybrid fibers with mixed solid–liquid mechanical properties. Proc Natl Acad Sci U S A 113(22):6143–6147

Foelix RF (2011) Biology of spiders, 3rd edn. Oxford University Press, Oxford

Foelix RF, Rast B, Peattie AM (2012) Silk secretion from tarantula feet revisited: alleged spigots are probably chemoreceptors. J Exp Biol 215(7):1084–1089

Foelix R, Erb B, Rast B (2013) Alleged silk spigots on tarantula feet: Electron microscopy reveals sensory innervation, no silk. Arthropod Struct Dev 42(3):209–217

Gerson U (1979) Silk production in *Tetranychus* (Acari: Tetranychidae). Recent Adv Acarol 1:177–188

Giribet G, Edgecombe GD, Wheeler WC, Babbitt C (2002) Phylogeny and systematic position of opiliones: a combined analysis of chelicerate relationships using morphological and molecular data. Cladistics 18(1):5–70

Glatz L (1973) Der Spinnapparat der Orthognatha (Arachnida, Araneae). Z Morphol Tiere 75(1):1–50

Gnaspini P, Hara MR (2007) Defense mechanisms. In: Pinto-da-Rocha R, Machado G, Giribet G (eds) Harvestmen: the biology of Opiliones. Harvard University Press, Cambridge, pp 374–399

Gorb SN, Niederegger S, Hayashi CY, Summers AP, Vötsch W, Walther P (2006) Biomaterials: silk-like secretion from tarantula feet. Nature 443(7110):407–407

Gregson JD (1960) Morphology and functioning of the mouthparts of *Dermacentor andersoni* Stiles. Acta Trop 17(1):48–49

Grimm U (1985) Die Gnaphosidae Mitteleuropas (Arachnida, Araneae). Abh naturwiss Ver Hamburg, (NF) 26:1–318

Gruber J (1970) Die „*Nemastoma*"-Arten Nordamerikas (Ischyropsalididae, Opiliones, Arachnida). Ann Naturhist Mus Wien 74:129–144

Gruber J (1993) Beobachtungen zur Ökologie und Biologie von Dicranolasma scabrum (HERBST) (Arachnida: Opiliones) Teil I. Ann Naturhist Mus Wien 94:393–426

Hajer J, Hrubá L (2007) Wrap attack of the spider *Achaearanea tepidariorum* (Araneae: Theridiidae) by preying on mealybugs *Planococcus citri* (Homoptera: Pseudococcidae). J Ethol 25(1):9–20

Hazan A, Gertler A, Tahori A, Gerson U (1975) Spider mite webbing—III. Solubilization and amino acid composition of the silk protein. Comp Biochem Physiol B 51(4):457–462

Helbig R, Nickerl J, Neinhuis C, Werner C (2011) Smart skin patterns protect springtails. Plos One 6(9):e25105

Hennebert E, Viville P, Lazzaroni R, Flammang P (2008) Micro-and nanostructure of the adhesive material secreted by the tube feet of the sea star *Asterias rubens*. J Struct Biol 164(1):108–118

Hermann HR, Blum MS (1981) Defensive mechanisms in the social Hymenoptera. Soc Insects 2:77–197

Huang Y, Wang Y, Sun L, Agrawal R, Zhang M (2015) Sundew adhesive: a naturally occurring hydrogel. J R Soc Interface 12:20150226

Japyassú HF, Macagnan CR (2004) Fishing for prey: the evolution of a new predatory tactic among spiders (Araneae, Pholcidae). Revista de Etologia 6(2):79–94

Juberthie C (1964) Recherches sur la biologie des Opilions. Ann Spéléol 19:5–238

Kanazawa M, Sahara K, Saito Y (2011) Silk threads function as an 'adhesive cleaner'for nest space in a social spider mite. Proc R Soc B 278(1712):1653–1660
Kästner A (1941) 2. Ordnung der Arachnida: Pedipalpi Latreille = Geißel-Scorpione. In: Kükenthal W, Krumbach T (eds) Handbuch der Zoologie: chelicerata, vol 3. Walter de Gruyter & Co., Berlin, pp 1–76
Kemp D, Stone B, Binnington K (1982) Tick attachment and feeding: role of the mouthparts, feeding apparatus, salivary gland secretions, and the host response. In: Obenchain FD, Galun R (eds) Physiology of ticks, vol 1. Pergamon Press, Oxford, pp 119–167
Kinloch AJ (1987) Adhesion and adhesives: science and technology. Springer Science & Business Media, London
Kovoor J (1987) Comparative structure and histochemistry of silk-producing organs in arachnids. In: Nentwig W (ed) Ecophysiology of spiders. Springer, Berlin, pp 160–186
Kovoor J, Zylberberg L (1972) Histologie et infrastructure de la glande chélicérienne de *Scytodes delicatula* Sim. (Araneidae, Scytodidae). Ann Sci Nat Zool Paris 14:333–388
Kovoor J, Zylberberg L (1980) Fine-structural aspects of silk secretion in a spider (*Araneus diadematus*).1. Elaboration in the pyriform glands. Tissue Cell 12(3):547–556
Lemos F, Sarmento RA, Pallini A, Dias CR, Sabelis MW, Janssen A (2010) Spider mite web mediates anti-predator behaviour. Exp Appl Acarol 52(1):1–10
Lopez A (1987) Glandular aspects of sexual biology. In: Nentwig W (ed) Ecophysiology of spiders. Springer, Berlin, pp 121–132
Lopez A, Emerit M, Rambla M (1980) Contribution a l'étude de *Sabacon paradoxum* Simon 1879 (Opiliones, Palpatores, Ischyropsalididae). Stations nouvelles, particularités électromicroscopiques du prosoma et de ses appendices. In: Comptes Rendus de la 5è Colloque d'Arachnologie. pp 147–161
Martens J (1993) Further cases of paternal care in Opiliones (Arachnida). Trop Zool 6(1):97–107
McAlister W (1960) The spitting habit in the spider *Scytodes intricata* Banks (Family Scytodidae). Tex J Sci 12:17–20
Nachtigall W (1974) Biological mechanisms of attachment. Springer, Berlin
Naldrett MJ (1993) The importance of sulphur cross-links and hydrophobic interactions in the polymerization of barnacle cement. J Mar Biol Assoc UK 73(03):689–702
Neinhuis C, Barthlott W (1997) Characterization and distribution of water-repellent, self-cleaning plant surfaces. Ann Bot 79(6):667–677
Opell BD, Schwend HS, Vito ST (2011) Constraints on the adhesion of viscous threads spun by orb-weaving spiders: the tensile strength of glycoprotein glue exceeds its adhesion. J Exp Biol 214(13):2237–2241
Palmer JM, Coyle FA, Harrison FW (1982) Structure and cytochemistry of the silk glands of the mygalomorph spider Antrodiaetus unicolor (Araneae, Antrodiaetidae). J Morphol 174(3):269–274
Peattie AM, Dirks JH, Henriques S, Federle W (2011) Arachnids secrete a fluid over their adhesive pads. PLoS One 6(5), e20485
Pocock R (1895) Some suggestions on the origin and evolution of web-spinning in spiders. Nature 51:417–420
Polis GA, Sissom W (1990) Life history. In: Polis GA (ed) The biology of scorpions. Standford University Press, Stanford, pp 161–223
Raspotnig G, Leis H-J (2009) Wearing a raincoat: exocrine secretions contain anti-wetting agents in the oribatid mite, *Liacarus subterraneus* (Acari: Oribatida). Exp Appl Acarol 47(3):179–190
Requena GS, Buzatto BA, Munguía-Steyer R, Machado G (2009) Efficiency of uniparental male and female care against egg predators in two closely related syntopic harvestmen. Anim Behav 78(5):1169–1176
Rimsky-Korsakow AP (1924) Die Kugelhaare von *Nemastoma lugubre*, Müll. Zool Anz 60:1–16
Rind FC, Birkett CL, Duncan BJA, Ranken AJ (2011) Tarantulas cling to smooth vertical surfaces by secreting silk from their feet. J Exp Biol 214(11):1874–1879

Röper H (1977) Analytische Untersuchungen des Wehrsekretes von *Peripatopsis moseleyi* (Onychophora). Z Naturforsch 32:57–60

Rost K, Schauer R (1977) Physical and chemical properties of the mucin secreted by *Drosera capensis*. Phytochemistry 16(9):1365–1368

Roewer CF (1923) Die Weberknechte der Erde. Gustav Fischer, Jena

Sahni V, Blackledge TA, Dhinojwala A (2010) Viscoelastic solids explain spider web stickiness. Nat Commun 1:19

Sahni V, Blackledge TA, Dhinojwala A (2011) Changes in the adhesive properties of spider aggregate glue during the evolution of cobwebs. Sci Rep 1:41

Sahni V, Dhinojwala A, Opell BD, Blackledge TA (2014a) Prey capture adhesives produced by orb-weaving spiders. In: Asakura T, Miller T (eds) Biotechnology of silk. Springer Science + Business Media, Dordrecht, pp 203–217

Sahni V, Miyoshi T, Chen K, Jain D, Blamires SJ, Blackledge TA, Dhinojwala A (2014b) Direct solvation of glycoproteins by salts in spider silk glues enhances adhesion and helps to explain the evolution of modern spider orb webs. Biomacromolecules 15(4):1225–1232

Schaider M, Raspotnig G (2009) Unusual organization of scent glands in *Trogulus tricarinatus* (Opiliones, Trogulidae): evidence for a non-defensive role. J Arachnol 37(1):78–83

Schönhofer AL (2013) A taxonomic catalogue of the Dyspnoi Hansen and Sørensen, 1904 (Arachnida: Opiliones). Zootaxa 3679(1):1–68

Schwangart F (1907) Beiträge zur Morphologie und Systematik der Opilioniden: 1. Über das Integument der Troguloidae. Zool Anz 31:161–183

Seniczak S, Seniczak A (2013) Morphology of juvenile stages and ontogeny of three species of Damaeidae (Acari: Oribatida). Int J Acarol 39(2):160–179

Shear WA (1986) A cladistic analysis of the opilionid superfamily Ischyropsalidoidea, with descriptions of the new family Ceratolasmatidae, the new genus *Acuclavella*, and four new species. Am Mus Novit 2844

Shultz JW (1987) The origin of the spinning apparatus in spiders. Biol Rev 62(2):89–113

Silverman HG, Roberto FF (2007) Understanding marine mussel adhesion. Mar Biotechnol 9(6):661–681

Stanley E (2011) Egg hiding in four harvestman species from Uruguay (Opiliones: Gonyleptidae). J Arachnol 39(3):495–496

Stubbs D, Tillinghast EK, Townley MA, Cherim N (1992) Fibrous composite structure in a spider silk. Naturwissenschaften 79(5):231–234

Suter RB, Stratton GE (2009) Spitting performance parameters and their biomechanical implications in the spitting spider, *Scytodes thoracica*. J Insect Sci 9(1):62

Torres FG, Troncoso OP, Cavalie F (2014) Physical characterization of the liquid adhesive from orb-weaving spiders. Mat Sci Eng C-Mater 34:341–344

Townley MA, Tillinghast EK (2003) On the use of ampullate gland silks by wolf spiders (Araneae, Lycosidae) for attaching the egg sac to the spinnerets and a proposal for defining nubbins and tartipores. J Arachnol 31(2):209–245

Uhl G, Nessler SH, Schneider JM (2010) Securing paternity in spiders? A review on occurrence and effects of mating plugs and male genital mutilation. Genetica 138(1):75–104

Uhl G, Kunz K, Vöcking O, Lipke E (2014) A spider mating plug: origin and constraints of production. Biol J Linn Soc 113(2):345–354

Vetter RS (1980) Defensive behavior of the black widow spider *Latrodectus hesperus* (Araneae: Theridiidae). Behav Ecol Sociobiol 7(3):187–193

Voetsch W, Nicholson G, Müller R, Stierhof Y-D, Gorb S, Schwarz U (2002) Chemical composition of the attachment pad secretion of the locust Locusta migratoria. Insect Biochem Mol Biol 32:1605–1613

Voigt D, Gorb SN (2010) Egg attachment of the asparagus beetle *Crioceris asparagi* to the crystalline waxy surface of *Asparagus officinalis*. Proc R Soc B 277(1683):895–903

Vollrath F, Tillinghast E (1991) Glycoprotein glue beneath a spider web's aqueous coat. Naturwissenschaften 78(12):557–559

Vollrath F, Fairbrother WJ, Williams RJP, Tillinghast EK, Bernstein DT, Gallagher KS, Townley MA (1990) Compounds in the droplets of the orb spiders viscid spiral. Nature 345:526–528
Wachmann E (1970) Der Feinbau der sog. Kugelhaare der Fadenkanker (Opiliones, Nemastomatidae). Z Zellforsch 103:518–525
Weygoldt P (1966) Spermatophore web formation in a pseudoscorpion. Science 153(3744):1647–1649
Weygoldt P (1968) Vergleichend-embryologische Untersuchungen an Pseudoscorpionen. Z Morphol Tiere 63(2):111–154
Weygoldt P (1969) Biology of pseudoscorpions. Harvard University Press, Cambridge
Weygoldt P (2000) Whip spiders (Chelicerata: Amblypygi): their biology, morphology and systematics. Apollo Books, Stenstrup
Weygoldt P, Huber S (2013) Sperm transfer and maternal care in *Thelyphonus* cf *caudatus* from Lombok, Indonesia (Arachnida, Uropygi, Thelyphonida). Zool Anz 252(3):348–349
Willemart RH (2001) Egg covering behavior of the neotropical harvestman *Promitobates ornatus* (Opiliones, Gonyleptidae). J Arachnol 29(2):249–252
Witaliński W, Żuwała K (1981) Ultrastructural studies of egg envelopes in harvestmen (Chelicerata, Opiliones). Int J Invertebr Rep Dev 4(2):95–106
Wolff JO, Schönhofer AL, Schaber CF, Gorb SN (2014) Gluing the 'unwettable': soil-dwelling harvestmen use viscoelastic fluids for capturing springtails. J Exp Biol 217(19):3535–3544
Wolff JO, Grawe I, Wirth M, Karstedt A, Gorb SN (2015) Spider's super-glue: thread anchors are composite adhesives with synergistic hierarchical organization. Soft Matter 11(12):2394–2403
Wolff JO, García-Hernández S, Gorb SN (2016a) Adhesive secretions in harvestmen (Arachnida: Opiliones). In: Smith AM, Callows JA (eds) Biological adhesives. Springer, Berlin
Wolff JO, Schönhofer AL, Martens J, Wijnhoven H, Taylor CK, Gorb SN (2016b) The evolution of pedipalps and glandular hairs as predatory devices in harvestmen (Arachnida, Opiliones). Zool J Linn Soc 177:558–601

Chapter 9
Biological Functions and Evolutionary Aspects

Abstract This book demonstrates that there is an enormous diversity of attachment devices and adhesives among arachnids. This diversity is comparable to what has previously been found in insects. In this chapter, we discuss the evolutionary driving forces that might have been acted on both the occurrence and shaping of such structures and secretions, in relation to their biological functions. We discuss instances, where the evolution of such structures might have been driven by another precursor function. Examples for this are the locomotory attachment pads of spiders, in which the original function might have been prey capture, or those of amblypygids, where the original function was obviously attachment to the mother cuticle. We discuss the role of sexual selection, as well as herbivore-plant, predator-prey and parasite-host interactions in the evolution of attachment devices and adhesive secretions in arachnids. Further, we discuss the role of attachment devices in dispersal and microhabitat access.

Our survey of attachment devices and adhesives in the previous chapters elucidates their diversity in arachnids. Some of them may be multi-functional, others serve very specific purposes. In order to fulfill the same biological function, different mechanisms have been often evolved in parallel, based either on mechanical interlocking (structure-based), dry adhesion (van der Waals forces), or wet adhesion (capillary or viscous forces). Locomotory attachment pads, prey capture threads of spiders, mating plugs and camouflaging with soil particles represent good examples where different solutions have independently evolved for a similar kind of problem.

In the following, we will discuss factors that may have caused the evolution of attachment organs, and influenced their shaping. In order to understand the relationship between morphology and function, it should always be kept in mind that derived structures (apomorphies) can only evolve on the basis of precursor structures (plesiomorphies), and this, in turn, limits the possibilities of appearance of novelties (phylogenetic burden). Also the initial function of a structure is often different from the actual one (functional shift). The comparison with analogies (homoplasies) in very distantly related taxa, exhibiting a different body plan (like vertebrates), can help to understand the physical principles that favoured evolution of particular shapes or/and properties of structures.

J.O. Wolff, S.N. Gorb, *Attachment Structures and Adhesive Secretions in Arachnids*, Biologically-Inspired Systems 7,
DOI 10.1007/978-3-319-45713-0_9

9.1 Ontogeny and the Role of Maternal Care

In many arachnid orders showing maternal care, attachment pads are present in the prenymphal instar (Wolff et al. 2015a). In these instances, the attachment organs obviously specifically evolved to attach to the mother. The structure of prenymphal pads is rather simple. It is more or less a fluid filled bag with a non-sclerotized, softened cuticle. The softening of the membrane that contacts the substrate is achieved by a loosening of the underlying cuticular material, presumably by the local lack of some matrix proteins and very low degree of sclerotisation. This does not seem to demand lots of evolutionary steps, and may evolve rather quickly. Nonetheless, this structure seems to be a rather efficient principle to provide both high compliance and stability, since it is also frequently found in convergently evolved smooth adhesive foot pads of insects. At least, in the whip-spiders, the prenymphal pads can be regarded as the precursors of the locomotory pads of the subsequent free living instars. Hence, maternal care could have started up the evolution of attachment structures for locomotion.

9.2 Dispersal

Dispersal can be passive, by getting drifted by air or water currents. In these cases attachment is not demanded, except when reaching the new microhabitat site to settle there. Another strategy is phoresy, which means the transport using another, much larger, animal. Although this is a mode of passive dislocation, it may demand a proper attachment. In many Acaridida and Gamasida, there is a calyptostatic (pupa-like) deutonymph, the so-called hypopus, which attaches to phoretic hosts, in order to reach new habitats (Alberti and Coons 1999). Various specific attachment organs have evolved for this purpose, like claspers, suckers or sticky secretions (Alberti and Coons 1999). Phoresy is also known from pseudoscorpions, which attach to fly legs with means of their pedipalp chelae (Zeh and Zeh 1992; Carl 1994). In this case, a multifunctional organ is used, which has unlikely evolved as just an adaptation to phoresy.

Depending on the size of the phoret, the host may encounter significant costs due to an increased weight, drag, or even injury (Carl 1994). Hence, host counter-adaptations, which reduce the attachment ability of the phoret, are expected, if there is a frequent and/or specific phoretic relationship. This issue is highly overlooked in the studies of phoresy. The abundance of phoretic mites on honeybees is significantly reduced by grooming behaviour (Delfinado-Baker et al. 1992), but a structure-based reduction of the attachment ability is not known. In case of the pseudoscorpion-fly relationship, the fly tries to remove the phoret by rubbing its legs together, but it is hardly able to remove it by this means, since pseudoscorpions can actively increase the clamping force (Carl 1994).

9.3 Microhabitat Access and Climbing Demand

Adhesive foot pads are widely used in locomotion, and associated with benefits, as well as costs. Phylogenetic analyses on geckoes (Gamble et al. 2012), insects (Beutel and Gorb 2001; Haas and Gorb 2004), and spiders (Wolff et al. 2013) revealed that the evolution of adhesive foot pads, their gain and loss, is highly dynamic. Adhesive foot pads are usually intuitively thought to enhance the climbing ability of the animal, which, however, is tested only in few exemplary studies in relation to the natural habitat and ecology of the animals. In some lizards (Zani 2000; Macrini et al. 2003; Elstrott and Irschick 2004), ants (Orivel et al. 2001), stick insects (Gottardo et al. 2015), and large ctenid spiders (Lapinski et al. 2015), a correlation between the inhabited stratum and the presence or size of adhesive pads was found. In lizards, there is also a correlation between claw structure and the height of the preferred microhabitat (Zani 2000). Such observations are often explained with the climbing demand, which is assumed to be much higher in arboreal microhabitats.

However, a plain relationship between arboreal living and the possession of adhesive pads or enlarged claws is not a general rule. All examples above comprise relatively large animals, with an exception of ants. For most arachnids, an adaptation of attachment organs to the microhabitat is not as obvious. In hunting spiders, for instance, claw tufts are found both in arboreal species and ones that strictly live on the ground. Further, there are adhesive pads in species of whip spiders, pseudoscorpions and ricinuleids that have exceptionally been found at the ground. On the other hand, there are numerous arboreal spiders that lack the adhesive foot pads, which is also true for most arboreal harvestmen and some climbing whip spiders.

The analysis of hunting spider assemblages revealed that the relative abundance of pad-bearing species increases with the height above the ground and the structuring of the micro-environment, which may indicate the differences in the ecological benefit of the pads (Wolff and Gorb 2014). However, web building spiders extremely rarely have adhesive pads (Wolff et al. 2013), although being often dominant in above-ground habitats. This shows that the mode of living and the associated demand on locomotion can play a much larger role in the evolution of locomotory attachment pads than the microhabitat niche. In order to understand the evolution of adhesive foot pads, it would be necessary to study the functional role and efficacy of claws more in detail. Presumably, most natural surfaces provide enough possibilities for interlocking, but it may take time to find a sufficient foothold, which increases the risk of attachment failure at a faster mode of locomotion. Therefore, it is possible that adhesive foot pads do not provide the benefit of a better general climbing ability (the ability to reach and dwell elevated sites), but enhance surefootedness, hence permitting an increased climbing speed and manoeuvrability in structured microhabitats. This might be the more beneficial, the higher the risk of dislodgment from the preferred site (i.e. by wind), the risk of capture by a predator, or, in the case of cursorial hunters, the faster the prey is.

Altogether, this shows that the evolution of locomotory attachment structures remains largely unclear, because it depends on a wide variety of factors. Keeping this in mind, it seems to be even more difficult to interpret, whether specific structural features of pads are adaptations to particular microhabitat surfaces. Natural surface structures are often very unpredictable and diverse in their properties, including topography, surface energy and amount of adsorbed water. All these properties highly influence adhesion, and thus attachment structures should be as universal as possible rather than show too narrow adaptations. Accordingly, in spider hairy adhesive pads, the structure and shape of the spatulate contact depends on their phylogenetic origin, not the microhabitat preferences of the spider species (Wolff et al. 2013).

Nonetheless, at least in very small arthropods with narrow niches, the observation of direct adaptation of the pretarsal attachment structures to the micro-environment seems probable. Based on their observations on intertidal mites, Pugh et al. (1987) argued that the pretarsal attachment structures are highly adaptive and that their structure reflects their ecological niche. In oribatids, attachment is primarily fulfilled by the action of claws (Heethoff and Koerner 2007), but species found in the canopy of rain forests exhibit arolia in the hind legs, which is presumed to be an essential adaptation to settle on the tree leaves (Walter and Behan-Pelletier 1993). In specialized herbivores and parasites, both of which among arachnids are only found in mites, stronger adaptation of attachment structures is more possible, as in these cases the substrates for attachment are more specific and less variable. Since these substrates are surfaces of organisms, which can equally evolve and have a direct ecological relationship to the attaching animal, there might be an antagonistic co-evolution (see next sections).

9.4 Herbivore-Plant Interaction

Despite of some clades of mites, arachnids are not herbivorous, but yet they are confronted with the evolutionary dynamics between herbivorous insects and plants. Many plants exhibit anti-adhesive waxy surfaces that may prevent the attachment of herbivorous insects or their eggs (Gorb 2001; Gorb and Gorb 2002, 2013; Eigenbrode 2004; Gorb et al. 2008; Whitney et al. 2009; Voigt and Gorb 2010; Whitney and Federle 2013). Arachnids that settle on plants may have to enhance the efficacy of their adhesive devices to keep a sufficient safety factor. Spiders use attachment discs of piriform gland silk to attach their safety lines and webs to substrates. The adhesion of the attachment discs is reduced on leaf surfaces, if compared with a smooth glass surface (Grawe et al. 2014). The arising question is, whether spiders can sense the strength of their attachment discs and either choose the best attachment sites or adjust the size or/and quality of the attachment discs. In *Eustala* spp. (Araneidae), it was found that dried dead plant stems were preferred as web attachment sites, which was explained with better camouflage (da Silva Souza et al. 2015). Dead or old leaves or stems often exhibit a reduced or damaged wax layer and an increased roughness by senescence (Mechaber et al. 1996; Neinhuis and Barthlott 1998), which may facilitate gluing onto these surfaces. However, whether the choice of foraging or web attachment sites is influenced by the adhesive capacity of the substrate surfaces has not been scope of a study yet. Nonetheless, it can be assumed that there is a high selective pressure on the glue efficacy. A previous study on beetle eggs that are glued on the usually highly anti-adhesive waxy surface of *Asparagus officinalis* came to the conclusion that the ability of the glue to wet the hydrophobic wax crystalloids is the key for the successful cementation of the eggs (Voigt and Gorb 2010). In spider attachment discs we found that the wax crystalloids partly contaminated the piriform gland glue leading to a reduction of attachment disc efficacy. However, the hierarchical structure of the disc reduces the local stress on the cementation interface, which increases the overall attachment performance (Wolff et al. 2015b). This may be an adaptation to enhance the safety of the thread anchors despite of the strong reduction of adhesion.

9.5 Predator-Prey Interaction

Most arachnids are predators. Predation demands morphological, physiological and behavioural means of grasping and arresting prey. In the present survey, we demonstrated the frequent occurrence of attachment devices or sticky secretions specialised for prey capture. Raptorial appendages are present in various clades and represent obvious adaptations to catch prey.

In spiders, the evolution of hairy adhesive pads highly correlates with the loss of prey capture web building behaviour (Wolff et al. 2013). We have hypothesized that the locomotory pads, the claw tufts, are secondary developments, and hence the effective prey capture function was the initial selective benefit that leads to the evolution and maintenance of the pads in spiders (Wolff et al. 2013). The pedipalpal smooth adhesive pads of solifuges are important means of prey capture, but they also serve other functions, such as climbing. However, it can be assumed that prey capture is the primary function of these organs, as climbing is rarely performed by these soil dwelling organisms.

In specific predator-prey relationships, there is a great selective pressure on the prey, forcing the evolution of adaptations against being caught (Dawkins and Krebs 1979; Vermeij 1994; Abrams 2000). Previous authors discussed the role of detachable setae or fluid coatings against the entrapment in spider webs (Nentwig 1982; Opell 1994; Opell and Schwend 2007). As a counter-adaptation against prey loss, some spiders evolved unique behaviours, like the release of elastic energy in webs that are hold pre-stressed (Uloboridae: *Hyptiotes*) (Marples and Marples 1937), net-casting (Deinopidae: *Deinopis*) (Getty and Coyle 1996) or the fishing with a single sticky line (Araneidae: *Mastophora*) (Eberhard 1980). Soil-dwelling harvestmen use viscoelastic glue to capture springtails (Wolff et al. 2014). Epigaeic springtails have evolved easily detachable scale-like setae, presumably as an anti-trapping device (Bauer and Pfeiffer 1991; Zhang et al. 2014; Wolff et al. 2014). Nemastomatid and sabaconid harvestmen have evolved hyper-bendable segments for clamping to compensate the loss of gluing efficacy (Wolff et al. 2016). The above examples show that adhesion can play an important role in an evolutionary 'arms-race' between preys and predators.

However, arachnids are also hunted by predators themselves. Attachment and gluing can be effectively used to deter or immobilize predators (Betz and Kölsch 2004), to attach to the ground to avoid being picked up, or for camouflage. Some spiders (Theridiidae and Pholcidae) use sticky threads for defence (Vetter 1980; Japyassú and Macagnan 2004). However, these probably did not specifically evolve for this purpose, but rather for prey immobilization. On the other hand, some structures or secretions specifically serve camouflage by catching soil particles: this phenomenon has independently evolved in some spiders, harvestmen and mites.

9.6 Parasite-Host Interaction

Parasites have an increased demand of attachment, since their life highly depends on their ability to keep their location on a specific host. The surfaces used for attachment may highly differ between host taxa, but may be uniform enough to evolve specific adaptations of parasites to attach to these surfaces. In astigmatid mites (Sarcoptiformes), the parasitic or phoretic deutonymphs show different types of attachment organs, which are related to the type of host: those that attach to arthropods have arolia, those that attach to mammalian hair have reduced arolia but claw-like setae, and those that burrow into the skin or live in hair follicles have all attachment structures reduced (Atyeo 1979; OConnor 1982; Fain 1994).

The generally accepted rules of parasite-host coevolution would assume the evolution of counter-measures against parasite attachment to the hosts. However, we could not find any published results showing evidence for a parasite-host coevolution of attachment and anti-attachment.

9.7 Sexual Selection

Females may favour short copulations with multiple partners to enhance sperm competition (cryptic female choice) (Parker 1970; Thornhill and Alcock 1983; Arnqvist and Nilsson 2000). In spiders and insects, it was shown that females try to interrupt copulation, whereas males increase their reproductive success with the duration of copulation (Thornhill and Alcock 1983; Elgar et al. 2000; Schneider et al. 2000). Hence, males may have a greater benefit, if they, or at least their copulatory structures, can strongly attach to the female. In insects, mate guarding frequently occurs as a strategy to secure paternity and avoid sperm concurrence (Alcock 1994). In these cases, males often have claspers to stay attached to the females (Gorb 2001; Willkommen et al. 2015). In male leaf beetles, the tarsal attachment pads exhibit specific setae that can generate higher adhesive forces on smooth surfaces, which is important to cling to the female's elytra (Stork 1983;

Pelletier and Smilowitz 1987; Voigt et al. 2008). In aquatic beetles, males attach to the female elytra using suction cups (Aiken and Khan 1992). Suction cups are rather susceptible to irregularities of the target surface. Interestingly, the female elytra exhibit a fine surface roughness, which reduces the efficacy of suction cups (Miller 2003; Green et al. 2013). Hence, it was assumed that this is an indication of the sexual conflict, which leads to a male-female co-evolution in an arms-race manner (Miller 2003; Bergsten and Miller 2007). Interestingly, acaridid mite males developed similar suction cups based devices to attach to the female, but, in contrast, the female seems to *improve* the attachment of the male in this case. The females often exhibit conspicuously smooth surfaces at the attachment site, and in some mites even the immature female possesses tubercles that perfectly fit into the modified tube-like suckers of the males (Witaliński et al. 1992). This contradicts the sexual conflict hypothesis, although also in females of acaridid mites multiple copulations frequently occur and are obviously advantageous (Witaliński et al. 1992). The reason for this phenomenon may be associated with the parasitic habit of these species, which means low densities of potential mating partners (Witaliński et al. 1992). If the period of receptivity of the female is short and the chance of male encounter is low it is beneficial to bind *any* found partner already before maturity, to ensure its presence, when the eggs have to be fertilized (Witaliński et al. 1992). Further, these structures may play a role in species recognition to avoid inter-specific copulations, which should enforce the evolution of compatibility instead of impairment as in the course of a sexual conflict. In general, specific attachment structures of males to attach to females, like clamping legs, are strikingly frequent among parasitic mites, and attachment often takes place already before the maturity of the female. The abundance of mating partners may thus play an important role in the evolution of mating-related attachment devices.

In the case of lock-and-key systems, like in entelegyne spiders, there might be an antagonistic co-evolution of the attachment ability. It is an exclusive mechanical interaction, since the copulatory structures usually do not bear sensory structures ('numb genitalia') (Eberhard and Huber 2010), with some rare exceptions (Lipke et al. 2015). It is barely possible to separate entelegyne spiders in copula without injuring the male (pers. observation on Gnaphosidae), which underlines the strength of the attachment. However, in females, selection may favour a reduction of the male attachment ability and hence an alteration of the structures, the male structures interlock with, in order to be able to interrupt and control the mating duration. This would increase the selective pressure on the interlocking ability of male pedipalpal structures. This hypothesis, explaining the often high complexity of pedipalps and vulvae, is highly neglected in the literature and should be studied more intensely in the future.

Spermatophores also often exhibit a complex capsule with protuberances. These structures are highly variable and species-specific, which leads to the presumption of a lock-and-key mechanism (Weygoldt 2000). However, the external genitals of the females are often rather simple and non-sclerotized (Peretti 2004). Hence, the structure may be shaped by mate choice in the way that females use the tactical cues given by the spermatophore structure to 'assess' the quality of the male (Peretti

2004). In a broad comparative approach, including both insects and arachnids, it was found that spermatophore complexity generally cannot be explained with a lock-and-key mechanism, but might rather be shaped by the female choice and the demand of protection of the sperm against environmental impacts (Proctor et al. 1995). Though, some of the protuberances often play a role in interlocking with the female genital tract, in order to increase the duration of sperm transfer (Weygoldt 2000; Jacob et al. 2004). This part of the interaction may be under the influence of sexual conflict.

Another strategy of males in securing paternity is the production of mating plugs, which seal the female genital opening (Parker 1970; Kuntner et al. 2009; Uhl et al. 2010). In contrast to insects, in arachnids this is much more common than mate guarding. It was shown that the efficacy of these plugs varies (Herberstein et al. 2012), which may be due to an antagonistic co-evolution of female and male structures and surfaces. However, such structures or surface coatings reducing the attachment ability of the plugs have not been identified yet. This may be explained by the frequent research focus on the male genitals alone (Ah-King et al. 2014). In some cases, plugs are produced by the females, even during copulation (Aisenberg and Barrantes 2011). The underlying biological function of this phenomenon is not entirely clear, but it is assumed that the female by this means ensures the paternity of her offspring by the partner of her choice (Aisenberg and Barrantes 2011). This implies that a rigorous mate choice is made before the copulation.

These examples illustrate that attachment structures and glues are frequent means of mate choice and securing paternity and may have specifically evolved under the pressure of sexual selection and sexual conflict. Further research is strongly required, in order to proof hypotheses shortly discussed in this chapter.

References

Abrams PA (2000) The evolution of predator-prey interactions: theory and evidence. Ann Rev Ecol Syst 31:79–105

Ah-King M, Barron AB, Herberstein ME (2014) Genital evolution: why are females still understudied? PLoS Biol 12(5):e1001851

Aiken R, Khan A (1992) The adhesive strength of the palettes of males of a boreal water beetle, *Dytiscus alaskanus* J. Balfour Browne (Coleoptera: Dytiscidae). Can J Zool 70(7):1321–1324

Aisenberg A, Barrantes G (2011) Sexual behavior, cannibalism, and mating plugs as sticky traps in the orb weaver spider *Leucauge argyra* (Tetragnathidae). Naturwissenschaften 98(7):605–613

Alberti G, Coons L (1999) Acari: Mites. In: Harrison FW, Foelix RF (eds) Microscopic anatomy of invertebrates: Chelicerate Arthropoda, vol 8. Wiley-Liss, New York, pp 515–1265

Alcock J (1994) Postinsemination associations between males and females in insects: the mate-guarding hypothesis. Annu Rev Entomol 39(1):1–21

Arnqvist G, Nilsson T (2000) The evolution of polyandry: multiple mating and female fitness in insects. Anim Behav 60(2):145–164

Atyeo WT (1979) Pretarsi of astigmatid mites. Acarologia 20:244–269

Bauer T, Pfeiffer M (1991) Shooting springtails with a sticky rod - The flexible hunting behavior of *Stenus comma* (Coleoptera, Staphylinidae) and the counterstrategies of its prey. Anim Behav 41:819–828

Bergsten J, Miller KB (2007) Phylogeny of diving beetles reveals a coevolutionary arms race between the sexes. Plos One 2(6):e522

Betz O, Kölsch G (2004) The role of adhesion in prey capture and predator defence in arthropods. Arthr Struct Dev 33(1):3–30

Beutel R, Gorb S (2001) Ultrastructure of attachment specializations of hexapods (Arthropoda): evolutionary patterns inferred from a revised ordinal phylogeny. J Zool Syst Evol Res 39(4):177–207

Carl M (1994) Injurious effects on the exoskeleton of *Musca domestica* L. (Diptera) of phoresy by *Lamprochernes nodosus* (Schrank) (Pseudoscorpiones, Chernetidae) and the possible functional significance of accessory teeth on the chelal fingers. Bull Brit Arachnol Soc 9(7):246–248

da Silva SH, Messas YF, de Oliveira GM, Vasconcellos-Neto J (2015) Substrate selection and spatial segregation by two congeneric species of *Eustala* (Araneae: Araneidae) in southeastern Brazil. J Arachnol 43(1):59–66

Dawkins R, Krebs JR (1979) Arms races between and within species. Proc R Soc B 205(1161):489–511

Delfinado-Baker M, Rath W, Boecking O (1992) Phoretic bee mites and honeybee grooming behavior. Int J Acarol 18(4):315–322

Eberhard WG (1980) The natural history and behavior of the bolas spider *Mastophora dizzydeani* sp. n. (Araneidae). Psyche 87(3–4):143–169

Eberhard WG, Huber BA (2010) Spider genitalia: precise maneuvers with a numb structure in a complex lock. In: Leonard JL, Cordoba-Aguilar A (eds) The evolution of primary sexual characters in animals. Oxford University Press, New York, pp 249–284

Eigenbrode SD (2004) The effects of plant epicuticular waxy blooms on attachment and effectiveness of predatory insects. Arthr Struct Dev 33(1):91–102

Elgar MA, Schneider JM, Herberstein ME (2000) Female control of paternity in the sexually cannibalistic spider *Argiope keyserlingi*. Proc R Soc B 267(1460):2439–2443

Elstrott J, Irschick DJ (2004) Evolutionary correlations among morphology, habitat use and clinging performance in Caribbean *Anolis* lizards. Biol J Linn Soc 83(3):389–398

Fain A (1994) Adaptation, specificity and host-parasite coevolution in mites (Acari). Int J Parasitol 24(8):1273–1283

Gamble T, Greenbaum E, Jackman TR, Russell AP, Bauer AM (2012) Repeated origin and loss of adhesive toepads in geckos. Plos One 7(6):e39429

Getty RM, Coyle FA (1996) Observations on prey capture and anti-predator behaviors of ogre-faced spiders (*Deinopis*) in southern Costa Rica (Araneae, Deinopidae). J Arachnol 24(2):93–100

Gorb S (2001) Attachment devices of insect cuticle. Springer Science & Business Media, Dordrecht

Gorb EV, Gorb SN (2002) Attachment ability of the beetle *Chrysolina fastuosa* on various plant surfaces. Ent Exp Appl 105:13–28

Gorb EV, Gorb SN (2013) Anti-adhesive surfaces in plants and their biomimetic potential. In: Fratzl P, Dunlop JWC, Weinkamer R (eds) RSC smart materials no. 4 materials design inspired by nature: function through inner architecture. The Royal Society of Chemistry, London, pp 282–308.

Gorb E, Voigt D, Eigenbrode SD, Gorb S (2008) Attachment force of the beetle *Cryptolaemus montrouzieri* (Coleoptera, Coccinellidae) on leaflet surfaces of mutants of the pea *Pisum sativum* (Fabaceae) with regular and reduced wax coverage. Arthr-Plant Inter 2(4):247–259

Gottardo M, Vallotto D, Beutel RG (2015) Giant stick insects reveal unique ontogenetic changes in biological attachment devices. Arthr Struct Dev 44(2):195–199

Grawe I, Wolff JO, Gorb SN (2014) Composition and substrate-dependent strength of the silken attachment discs in spiders. J R Soc Interface 11(98):1742–5662

Green KK, Kovalev A, Svensson EI, Gorb SN (2013) Male clasping ability, female polymorphism and sexual conflict: fine-scale elytral morphology as a sexually antagonistic adaptation in female diving beetles. J R Soc Interface 10(86):20130409

Haas F, Gorb S (2004) Evolution of locomotory attachment pads in the Dermaptera (Insecta). Arthr Struct Dev 33(1):45–66
Heethoff M, Koerner L (2007) Small but powerful: the oribatid mite *Archegozetes longisetosus* Aoki (Acari, Oribatida) produces disproportionately high forces. J Exp Biol 210(17):3036–3042
Herberstein M, Wignall A, Nessler S, Harmer A, Schneider J (2012) How effective and persistent are fragmentsof male genitalia as mating plugs? Behav Ecol 23(5):1140–1145
Jacob A, Gantenbein I, Braunwalder ME, Nentwig W, Kropf C (2004) Morphology and function of male genitalia (spermatophores) in *Euscorpius italicus* (Euscorpiidae, Scorpiones): Complex spermatophore structures enable safe sperm transfer. J Morphol 260(1):72–84
Japyassú HF, Macagnan CR (2004) Fishing for prey: the evolution of a new predatory tactic among spiders (Araneae, Pholcidae). Revista de Etologia 6(2):79–94
Kuntner M, Kralj-Fišer S, Schneider J, Li D (2009) Mate plugging via genital mutilation in nephilid spiders: an evolutionary hypothesis. J Zool 277(4):257–266
Lapinski W, Walther P, Tschapka M (2015) Morphology reflects microhabitat preferences in an assemblage of neotropical wandering spiders. Zoomorph 134(2):219–236
Lipke E, Hammel JU, Michalik P (2015) First evidence of neurons in the male copulatory organ of a spider (Arachnida, Araneae). Biol Lett 11(7):20150465
Macrini TE, Irschick DJ, Losos JB (2003) Ecomorphological differences in toepad characteristics between mainland and island anoles. J Herpetol 37(1):52–58
Marples MJ, Marples BJ (1937) Notes on the spiders *Hyptiotes paradoxus* and *Cyclosa conica*. Proc Zool Soc Lond 3:213–221
Mechaber WL, Marshall DB, Mechaber RA, Jobe RT, Chew FS (1996) Mapping leaf surface landscapes. Proc Natl Acad Sci U S A 93(10):4600–4603
Miller KB (2003) The phylogeny of diving beetles (Coleoptera: Dytiscidae) and the evolution of sexual conflict. Biol J Linn Soc 79(3):359–388
Neinhuis C, Barthlott W (1998) Seasonal changes of leaf surface contamination in beech, oak, and ginkgo in relation to leaf micromorphology and wettability. New Phytol 138(1):91–98
Nentwig W (1982) Why do only certain insects escape from a spider's web? Oecologia 53(3):412–417
O'Connor B (1982) Evolutionary ecology of astigmatid mites. Annu Rev Entomol 27(1):385–409
Opell B (1994) The ability of spider cribellar prey capture thread to hold insects with different surface features. Funct Ecol 8(2):145–150
Opell BD, Schwend HS (2007) The effect of insect surface features on the adhesion of viscous capture threads spun by orb-weaving spiders. J Exp Biol 210(13):2352–2360
Orivel J, Malherbe M, Dejean A (2001) Relationships between pretarsus morphology and arboreal life in ponerine ants of the genus *Pachycondyla* (Formicidae: Ponerinae). Ann Entomol Soc Am 94(3):449–456
Parker GA (1970) Sperm competition and its evolutionary consequences in the insects. Biol Rev 45(4):525–567
Pelletier Y, Smilowitz Z (1987) Specialized tarsal hairs on adult male Colorado potato beetles, *Leptinotarsa decemlineata* (Say), hamper its locomotion on smooth surfaces. Can Entomol 119(12):1139–1142
Peretti AV (2004) Functional morphology of spermatophores and female genitalia in bothriurid scorpions: genital courtship, coercion and other possible mechanisms. J Zool 261(02):135–153
Proctor HC, Baker RL, Gwynne DT (1995) Mating behaviour and spermatophore morphology: a comparative test of the female-choice hypothesis. Can J Zool 73(11):2010–2020
Pugh P, King P, Fordy M (1987) Ambulacral structure in the terrestrial moiety of the intertidal Acari, and its relationship with the lifestyle of the Acari. Acarologia 28(1):3–14
Schneider J, Herberstein M, De Crespigny FC, Ramamurthy S, Elgar M (2000) Sperm competition and small size advantage for males of the golden orb-web spider *Nephila edulis*. J Evol Biol 13(6):939–946

Stork NE (1983) The adherence of beetle tarsal setae to glass. J Nat Hist 17(4):583–597
Thornhill R, Alcock J (1983) The evolution of insect mating systems. Harvard University Press, Boston
Uhl G, Nessler SH, Schneider JM (2010) Securing paternity in spiders? A review on occurrence and effects of mating plugs and male genital mutilation. Genetica 138(1):75–104
Vermeij GJ (1994) The evolutionary interaction among species: selection, escalation, and coevolution. Ann Rev Ecol Syst 25:219–236
Vetter RS (1980) Defensive behavior of the black widow spider *Latrodectus hesperus* (Araneae: Theridiidae). Behav Ecol Sociobiol 7(3):187–193
Voigt D, Gorb SN (2010) Egg attachment of the asparagus beetle *Crioceris asparagi* to the crystalline waxy surface of *Asparagus officinalis*. Proc R Soc B 277:895–903
Voigt D, Schuppert JM, Dattinger S, Gorb SN (2008) Sexual dimorphism in the attachment ability of the Colorado potato beetle *Leptinotarsa decemlineata* (Coleoptera: Chrysomelidae) to rough substrates. J Insect Physiol 54:765–776
Walter DE, Behan-Pelletier VM (1993) Systematics and ecology of *Adhaesozetes polyphyllos* sp. nov. (Acari: Oribatida: Licneremaeoidea), a leaf-inhabiting mite from Australian rainforests. Can J Zool 71(5):1024–1040
Weygoldt P (2000) Whip spiders (Chelicerata: Amblypygi): their biology, morphology and systematics. Apollo Books, Stenstrup
Whitney HM, Federle W (2013) Biomechanics of plant-insect interactions. Curr Opin Plant Biol 16(1):105–111
Whitney H, Federle W, Glover B (2009) Grip and slip: mechanical interactions between insects and the epidermis of flowers and flower stalks. Integr Comp Biol 2(6):505–508
Willkommen J, Michels J, Gorb SN (2015) Functional morphology of the male caudal appendages of the damselfly *Ischnura elegans* (Zygoptera: Coenagrionidae). Arthr Struct Dev 44(4):289–300
Witaliński W, Dabert J, Walzl MG (1992) Morphological adaptation for precopulatory guarding in astigmatic mites (Acari: Acaridida). Int J Acarol 18(1):49–54
Wolff JO, Gorb SN (2014) Adhesive foot pads: an adaptation to climbing? An ecological survey in hunting spiders. Zoology 118:1–7
Wolff JO, Nentwig W, Gorb SN (2013) The great silk alternative: multiple co-evolution of web loss and sticky hairs in spiders. Plos One 8(5):e62682
Wolff JO, Schönhofer AL, Schaber CF, Gorb SN (2014) Gluing the 'unwettable': soil-dwelling harvestmen use viscoelastic fluids for capturing springtails. J Exp Biol 217(19):3535–3544
Wolff J, Huber S, Gorb S (2015a) How to stay on mummy's back: morphological and functional changes of the pretarsus in arachnid postembryonic stages. Arthr Struct Dev 44:301–312
Wolff JO, Grawe I, Wirth M, Karstedt A, Gorb SN (2015b) Spider's super-glue: thread anchors are composite adhesives with synergistic hierarchical organization. Soft Matter 11(12):2394–2403
Wolff JO, Martens J, Schönhofer AL, Gorb SN (2016) Evolution of hyper-flexible joints in sticky prey capture appendages of harvestmen (Arachnida, Opiliones). Organism Divers Evol, in press, online pre-pub
Zani P (2000) The comparative evolution of lizard claw and toe morphology and clinging performance. J Evol Biol 13(2):316–325
Zeh DW, Zeh JA (1992) Failed predation or transportation? Causes and consequences of phoretic behavior in the pseudoscorpion *Dinocheirus arizonensis* (Pseudoscorpionida: Chernetidae). J Insect Behav 5(1):37–49
Zhang F, Chen Z, Dong R-R, Deharveng L, Stevens MI, Huang Y-H, Zhu C-D (2014) Molecular phylogeny reveals independent origins of body scales in Entomobryidae (Hexapoda: Collembola). Mol Phylogenet Evol 70:231–239

Chapter 10
Comparative Contact Mechanics

Abstract In this chapter, we briefly discuss three hotly debated issues of the current research on adhesive foot pads, (1) the validity of the contact splitting theory and its implication for scaling laws, (2) the role of fluid secretions in hairy adhesive pads, and (3) the controversy about the subdivision of biological adhesive systems into 'dry' and 'wet' adhesives. We present the first data on comparative adhesion measurements in arachnids on (a) smooth pads, (b) patterned soft pads, and (c) hairy adhesive pads covered with tenent microtrichia. The data generally indicate stronger adhesion in systems with multiple contacts. Taking our new data on hairy adhesive pads in arachnids into account, we attempt to explain, why the evolution of spatular sizes is governed by phylogenetic constraints and the previously postulated relationship between body mass and degree of contact splitting does not hold. Furthermore, we provide several arguments against the view of previous authors, who classified the hairy adhesive pads of spiders and mites as wet adhesives, and stated geckoes as the sole organisms relying on dry adhesion.

10.1 Comparative Mechanics and Strength of Smooth and Structured Adhesive Foot Pads

Previous authors have speculated about which of the two structural principles, the smooth or the hairy pads, is more efficient and what this means for the evolution of locomotory attachment pads and the design of artificial adhesives (Gorb 2001; Arzt et al. 2003; Federle 2006; Bullock et al. 2008; Varenberg et al. 2011). We comparatively measured the vertical pull-off force of different arachnids from glass (see Chap. 1 for methodology). The data of pull-off force and adhesive strength are summarized in Fig. 10.1. Despite of the largest individuals of whip-spiders, the safety factor was high enough to support multiple times of the animal body weight in all species tested. The safety factor was comparable between taxa with smooth and those with hairy adhesive pads. However, in arachnids with smooth pads, a larger contact area (and presumably a higher amount of fluid) was observed. As a general rule, the efficacy of the pads depends both on the body size and the structure of the pad. The efficacy decreases with increasing body weight, due to the decreasing surface to volume ratio and presumably the decreasing evenness of load distribution

J.O. Wolff, S.N. Gorb, *Attachment Structures and Adhesive Secretions in Arachnids*, Biologically-Inspired Systems 7,
DOI 10.1007/978-3-319-45713-0_10

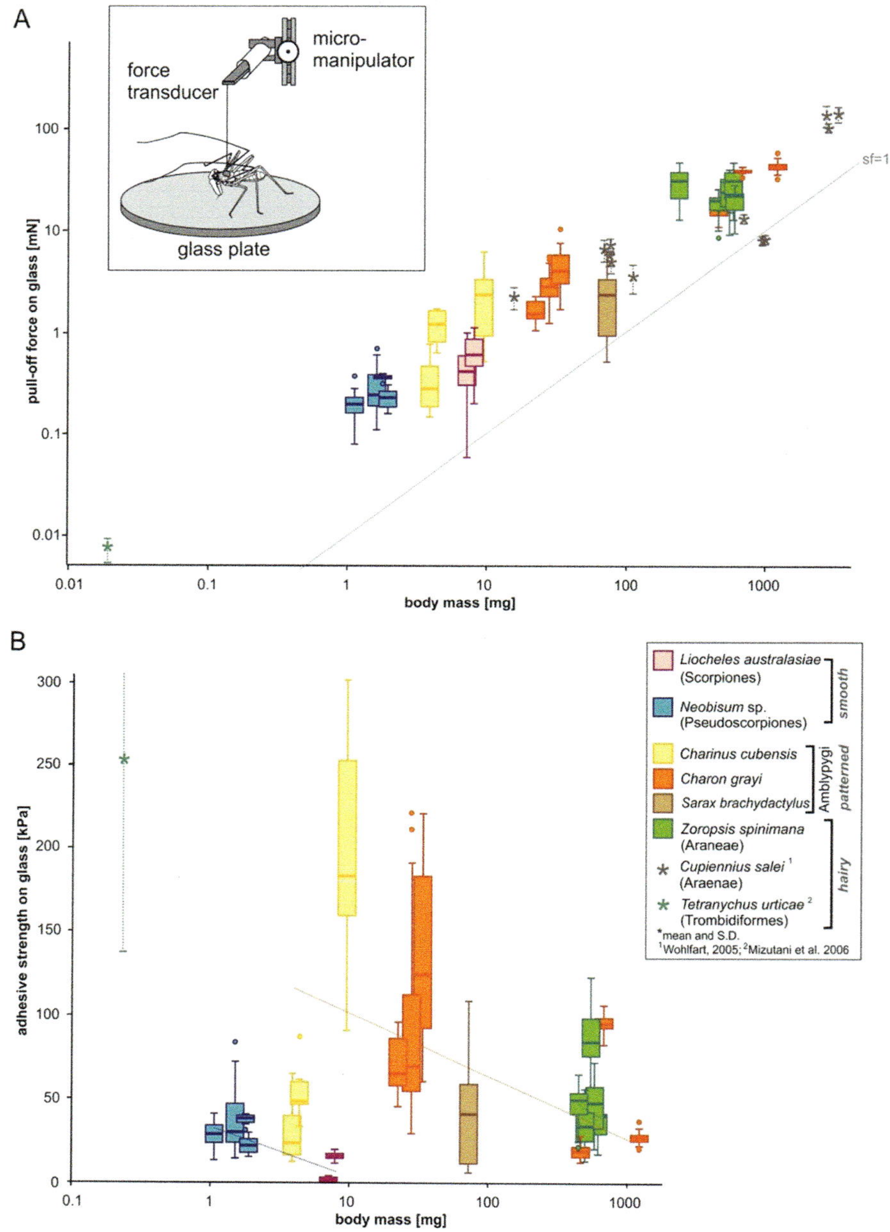

Fig. 10.1 Results of comparative force measurements on adhesive foot pads. Boxplots give the median value, interquartile range and the extreme values, outliers are indicated by circles. Each box plot illustrates the data obtained on a single individual (10–25 measurements on each). In *C. salei* (different individuals) and *T. urticae* (data on several individuals are pooled together) the mean and standard deviation are illustrated, data taken from Wohlfart (2005) and Mizutani et al. (2006). (**a**) Peak pull-off forces of animals attached to a smooth glass slide. The inset illustrates the measurement setup. (**b**) Adhesive strength (pull-off forces divided by the projected contact area). Data of *C. salei* is not included here, because the data on the contact area of the experimental animals was lacking. The adhesive strength decreases with increasing mass, because of the allometric relationship between surface and volume, and the presumably decreasing evenness of stress distribution in the contact area. Note the much lower values of smooth adhesive pads, indicating that a much higher proportion of the pad area has to be used to generate similar adhesive forces as in hairy and patterned adhesive pads

(see Labonte and Federle 2015b). Of course, one has to interpret these data with caution, because our comparison has been done on a smooth substrate, whereas in the natural situation, these systems should cope with rough surfaces, which strongly influence the real contact area between smooth/hairy pads and the substrate.

According to our data on arachnids, the animals with spatulate microstructures could gain much higher adhesive strength than those with smooth pads. There was no obvious difference in adhesive strength between hairy pads (spiders) and arolia with spatulae (whip-spiders). However, in whip-spiders, the performance highly differed between individuals, which might be due to different amounts or composition of aroliar secretion or differences in the physical condition of the animals. A single individual of *Charinus cubensis* could not generate significantly higher adhesive strength than the smooth padded *Neobisium* sp.. The *Sarax brachydactylus* individual generated a significantly lower adhesive strength than the *Charon* cf. *grayi* individuals and one individual of *C. cubensis. S. brachydactylus* lacks clear spatulae, but has only roof like microstructures (Wolff et al. 2015), which may have an effect on the adhesive performance under applied shear force. However, as only a single individual of this species was available for the experimental study, it is not clear whether this generally holds. Both tested *Liocheles australasiae* prenymphs generated much lower adhesive strength than *Neobisium* sp., which cannot be totally explained by the allometric effect. This may indicate a lower efficacy of this 'primitive' pad with its circular contact area.

10.2 The Contact Splitting Theory and the Scaling of Adhesive Foot Pads

In order to extract some general functional principles of locomotory attachment pads and detect their limits, scaling laws have to be taken into account. This is also of high importance, because observations on small animals are intended to be used in biomimetic, usually larger scale, applications. As discussed above, adhesion depends on the overall contact area. In tape-like contact elements, it depends on the total edge length (peeling line), and in case of fluid supplemented setae, small suckers or mushroom-like contact elements on the circumference of the contact. Because growing bodies roughly increase their volume a power higher than their surface area, the available pad area relative to the body mass decreases with an increasing body mass. Thus, the only way to achieve sufficient adhesion to hold such body mass on the wall or ceiling is an enhancement of the adhesion strength (adhesion force per pad area).

Previous authors argued that such an increase of efficacy can be achieved by splitting the total contact area into fine sub-contacts (Scherge and Gorb 2001; Arzt et al. 2003; Peressadko and Gorb 2004), because this may hamper the propagation of the delamination crack in the contact and increase the overall peeling length or sum of circumferences (Chung and Chaudhury 2005; Varenberg et al. 2010). This, so-called, contact-splitting theory predicts that the degree of splitting in different groups of organisms increases with body mass (Arzt et al. 2003). Accordingly, Arzt et al. (2003) observed in a broad comparative analysis of the density and size of sub-contacts that this relationship holds for animals with hairy adhesive pads from flies to geckoes.

However, Peattie and Full (2007) argued that this study was not corrected for phylogenetic dependence and that the the-larger-the-denser relationship does not hold within lineages that independently acquired the adhesive pads. Nonetheless, this view does not overall reject the contact splitting theory, because the different compared groups (insects, spiders, lizards) differ in their body size ranges and may therefore had different demands on the adhesive strengths of the pads throughout the evolution of their adhesive setae. Recently, Labonte et al. (2016) found that the increase in adhesion forces with body size is primarily attributed to the simple increase of pad area, regardless of the fact whether the pads are smooth or hairy. Furthermore, the authors argued that there is still no experimental evidence that the adhesion strength on smooth surfaces increases with contact splitting. In fact, a study that found significantly higher adhesion strength in a polymer model with a structured surface compared to a similar sized model with a smooth surface, based their calculations on the real contact area (Peressadko and Gorb 2004). In this case, the total adhesion force on a smooth substrate was even smaller in the structured sample. This was supported by experiments using different models, in which the real contact area was kept constant (Varenberg et al. 2011). In contrast, adhesion enhancement is clearly present in models with mushroom-like contact elements, despite of the reduced real contact area, if compared with a smooth surface control (Gorb et al. 2007). However, in this case it is difficult to clearly relate the enhancement effect to contact splitting, since the overall flexibility of the material is simultaneously altered by the introduction of microstructures. Comparative studies of attachment forces on insects did not show the presumed relationship between contact size and adhesive strength (Gorb et al. 2001; Bullock et al. 2008; Labonte and Federle 2015b).

Our data on arachnids now shed new light on these issues. Our comparative data on arachnids with smooth and hairy pads clearly indicate a positive effect of contact splitting on the adhesion strength. This is most striking when comparing pseudoscorpions and whip-spiders, which bear arolia with a very similar structure and function, but with either a smooth or a patterned contact area. However, the effect of difference in material properties and fluid properties cannot be completely excluded.

Underlining a beneficial effect of contact splitting, we observed the crack arresting effect of spatulate microstructured pads for the first time *in vivo*. RICM-HSV clearly show an even peel-off mechanism of smooth pads (Videos 7.6, 7.7, 7.15, 7.16 and 7.20), whereas the delamination crack propagates stepwise and discontinuous in the patterned pads of whip-spiders (Videos 7.11 and 7.13) and the hairy pads of spiders (Videos 3.2 and 3.8).

Nonetheless, this does not necessarily mean that larger animals bear finer subcontacts. Our data demonstrate, too, that animals are stuck in their phylogenetic constraints. For instance, larger whip-spiders cannot increase their adhesion strength by growing finer spatulae, because the size might be set by the material structure they are made of, or/and by conservative genetic programs that may only slowly change throughout evolution. In the whip-spiders, we observed that larger individuals must use a larger proportion of their pad for adhesion generation, and, although not quantified, we more frequently observed a viscous secretion in these, too (Wolff et al. 2015). Eventually, very large individuals of *Charongrayi* cannot hold their body weight anymore on a steep smooth glass surface (Wolff et al. 2015).

The phylogenetic burden is quite illustrative, when comparing the ultrastructure of the spatula of tenent microtrichia of araneomorph spiders and trombidiform mites. In both, the microtrichia are formed by a pair of cuticular fibrils that apically bifurcate, spanning up a delicate membrane of the interconnecting matrix. Due to the phylogenetic distance of such spiders and mites, it is clear that these structures evolved twice, but presumably gained their similar structure due to a same precursor structure, namely the specific fibrillary structure of the arachnid cuticle. In consequence, the size and density of spatulae in the adhesive pads are highly similar, although the body masses of mites and spiders range over several orders of magnitude. It should be added that the size and density of the spatulae in mites, which include the smallest animals with hairy adhesive pads, and spiders are similar to geckoes, which include the largest animals with hairy adhesive pads (Fig. 10.2). It must be concluded that the study of Arzt et al. (2003) has a strong bias of the selected taxa, for instance only two species of spiders were included, both of which are very large.

To sum up, our new data on arachnids (1) indicates that contact splitting enhances the adhesion strength, which has never been shown before experimentally for biological systems, but (2) that phylogenetic constraints strongly affect the evolution of spatula sizes. Evidence for the correlation between body mass and the degree of contact splitting is still missing for clearly homologous structures.

10.3 Dry or Wet Adhesion?

Biological adhesives are traditionally divided into wet and dry adhesives, to emphasize the relative role of the microstructure and the fluid. Truly dry adhesives are believed to be better understood and to be more easily reproducible by biomimetic manufacturing. However, it is not easy to clearly define and delimitate functional types of dry and wet adhesives, because often both structural and rheological effects act synergistically.

It was previously assumed that a thin homogeneous fluid film between the pad and the substrate is essential for good adhesion in all kinds of smooth adhesive foot pads (Dixon et al. 1990; Jiao et al. 2000; Drechsler and Federle 2006; Barnes 2007; but see Labonte and Federle 2015a for alternative view). In hairy adhesive foot pads, consisting of un-branched setae with spatulae, the latter are too large to sufficiently adapt to the particular micro- and nano-roughness of a substrate, and fluids may act as a bonding agent (Edwards and Tarkanian 1970; Stork 1980; Gorb 2001; Bullock et al. 2008; Dirks et al. 2009; Gorb et al. 2010) (Fig. 10.3). Occasionally,

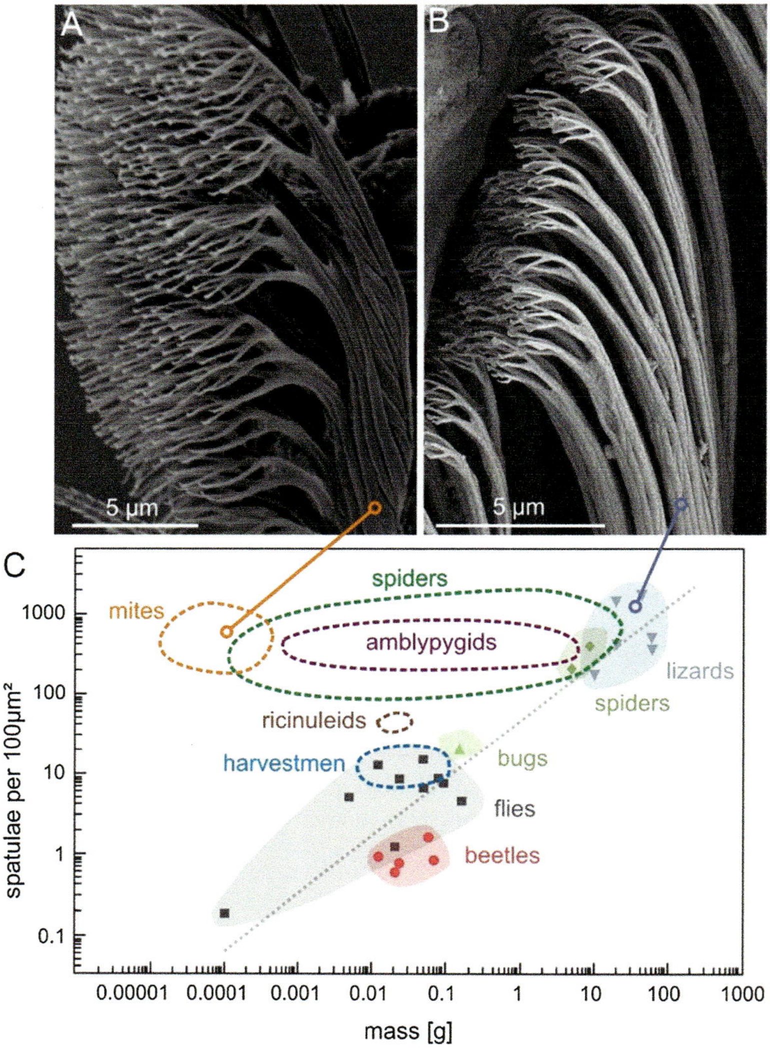

Fig. 10.2 Analogy and scaling of hairy adhesive foot pads. The structure of dry adhesive pads is remarkably consistent, as for example in the analogue setae of (**a**) trombidiform mites (here: *Anystis* sp.) and (**b**) lizards (here: *Gecko gekko*). (**c**) The mapping of our estimates for arachnid foot pads (shapes with *dashed lines*) on the plot of Arzt et al. (2003), shows that the hypothesized general scaling law of an increase of contact splitting with body mass (*grey dotted line*) does not hold

larger amounts of secretion emergence have been observed, which is attributed to the removal of contaminating dirt particles (Clemente et al. 2010).

In hairy pads, with a high density of very small spatulae, such as those in geckoes, spiders and mites, it was assumed that fluids barely play a role and the adhesion is primarily attributed to van der Waals forces (dry adhesion) (Autumn et al. 2002; Kesel et al. 2003; Mizutani et al. 2006). This means that the structure and material

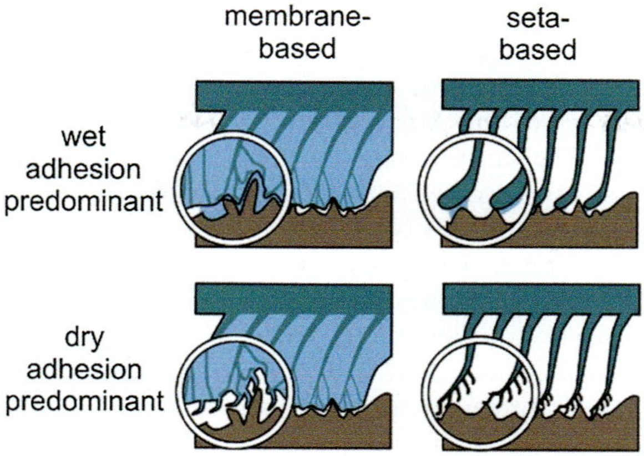

Fig. 10.3 Structural types of adhesive foot pads and their correlation with adhesive fluids. Although some fluid secretions may occur in all types of adhesive foot pads, both their occurrence and amount differ greatly. As a general rule, it is hypothesized that the presence of secretions in the contact zone is necessary for the generation of high adhesion in smooth pads and spatulate setae (*upper row*), but only play a minor role for adhesion generation in patterned pads and spatulate microtrichia of a submicron dimension (*bottom row*)

alone are responsible for good adhesion, which renders them most interesting for biomimetic applications (Autumn and Gravish 2008). However, this issue is still controversially discussed, as traces of secretions have been found on the spatulae of geckoes using very sensitive methods of analytical chemistry (Hsu et al. 2011). It was thus assumed that structural adhesive pads always imply some kind of secretion, but in very small spatulae, the amounts are usually too small to be seen in the microscope. Furthermore, at the nanoscale, adsorbed water molecules have a considerable effect on adhesion (Jang et al. 2004; Asay and Kim 2006). Small amounts of adsorbed water are usually present on natural surfaces, even in a dry environment (Israelachvili 1991). Experiments with geckoes and spiders show an effect of humidity on pad adhesion (Huber et al. 2005; Niewiarowski et al. 2008; Wolff and Gorb 2012), which may be attributed to the amount of adsorbed water on both substrate and spatula surfaces and hence to the capillary effect, as proposed by previous authors (Homann 1957; Roscoe and Walker 1991). Water may also be *ab*sorbed by the contact element, softening its material and leading to higher compliancy (Prowse et al. 2011; Peisker et al. 2013).

For the first time, Peattie et al. (2011) comparatively studied contact formation of adhesive pads of different arachnids to glass surfaces, revealing the occasional presence of fluid secretions. The authors used this finding to highlight the gecko adhesive pad as the only true dry adhesive system, an opinion, which was later taken up (Labonte and Federle 2015a). However, the authors neither did comparative tests with geckos, nor provided explanations how the arachnid secretions may support adhesion. The data presented by Peattie et al. (2011) for spiders show an extreme wetting of the setae, such that the spatulae are submerged in the fluid. Adhesion models show that in hairy pads, secretion enhances adhesion, only if it wets the space between the spatula and the substrate (Huber et al. 2005). If the fluid film is thicker than the spatula, slipping occurs (Bullock et al. 2008; Wolff and Gorb 2012). Like Peattie et al. (2011) we occasionally observed fluids in the adhesive pads of a palpimanid spider from South-Africa using RICM (Videos 3.6 and 3.7). In the cases of fluid appearance, the setal tips were completely wetted and the pad slid along the surface. Hence, adhesion was obviously reduced. Experiments with dead specimens of spider legs show that their hairy adhesive pads can generate high adhesion and friction on smooth surfaces in a dry state (Kesel et al. 2003; Niederegger and Gorb 2006; Wolff and Gorb 2013). Thus, the secretions observed by Peattie et al. (2011) and us unlikely support adhesion, but may have the function to clean the setae, control (reduce) adhesion, or may even result from unusual behavior of the animals for example due to the strong illumination used in such visualisation methods. Furthermore, lipid secretions are also present in the feet of web-building spiders, such as orb-web spiders (Araneidae) (Kropf et al. 2012), cob-web spiders (Theridiidae), and cellar spiders (Pholcidae) (own observation), all of which do not bear adhesive pads. In these spiders, the secretion prevents them from getting stuck in the own webs (Briceño and Eberhard 2012; Kropf et al. 2012). This shows that the foot secretions are not necessarily correlated with the presence of an adhesive pad and with adhesion enhancement.

Nonetheless, in trombidiform mites, which bear spatulate microtrichia comparable to spiders, there are single observations of fluids only wetting the contact element (see Chaps. 3 and 7, and Mizutani et al. 2006). By adhesion measurements on substrates with different free surface energy and theoretical modelling Mizutani et al. (2006) argued that fluids may only play a minor role in small contact elements like the spatulae of spiders and trombidiform mites.

In contrast, fluid secretions are generally regarded as essential for the adhesion of smooth adhesive pads (Dixon et al. 1990; Jiao et al. 2000; Drechsler and Federle 2006; Barnes 2007). We observed small amounts of lipoid secretions in the smooth arolia of pseudoscorpions, scorpion and whip scorpion nymphs, ticks, and anystid mites. However, the deposited amounts may vary greatly, and can even be totally absent (compare Videos 7.6, 7.7, 7.15, 7.16, 7.19, 7.20). Hence their functional significance for adhesion is unclear, and has recently even been overall questioned (Labonte and Federle 2015a). In amblypygids, which possess spatulae, we found that the pad is generating high adhesion forces in both a wet and a dry mode (Wolff et al. 2015) (Videos 7.11, 7.12, 7.13 and 7.14). This is the first known case of an arolium working according to the principle of dry adhesion, which is attributed to the particular surface microstructure. Interestingly, if present, the secretion emerges and assembles in the spaces between the spatulae (Videos 7.12 and 7.13), which should be disadvantageous for adhesion (see above), thus, its exact function remains obscure.

References

Arzt E, Gorb S, Spolenak R (2003) From micro to nano contacts in biological attachment devices. Proc Natl Acad Sci U S A 100(19):10603–10606

Asay DB, Kim SH (2006) Effects of adsorbed water layer structure on adhesion force of silicon oxide nanoasperity contact in humid ambient. J Chem Phys 124(17):174712
Autumn K, Gravish N (2008) Gecko adhesion: evolutionary nanotechnology. Phil Trans R Soc A 366(1870):1575–1590
Autumn K, Sitti M, Liang YA, Peattie AM, Hansen WR, Sponberg S, Kenny TW, Fearing R, Israelachvili JN, Full RJ (2002) Evidence for van der Waals adhesion in gecko setae. Proc Natl Acad Sci U S A 99(19):12252–12256
Barnes WJ (2007) Functional morphology and design constraints of smooth adhesive pads. MRS Bull 32(06):479–485
Briceño R, Eberhard W (2012) Spiders avoid sticking to their webs: clever leg movements, branched drip-tip setae, and anti-adhesive surfaces. Naturwissenschaften 99(4):337–341
Bullock JM, Drechsler P, Federle W (2008) Comparison of smooth and hairy attachment pads in insects: friction, adhesion and mechanisms for direction-dependence. J Exp Biol 211(20):3333–3343
Chung JY, Chaudhury MK (2005) Roles of discontinuities in bio-inspired adhesive pads. J R Soc Interface 2(2):55–61
Clemente CJ, Bullock JM, Beale A, Federle W (2010) Evidence for self-cleaning in fluid-based smooth and hairy adhesive systems of insects. J Exp Biol 213(4):635–642
Dirks J-H, Clemente CJ, Federle W (2009) Insect tricks: two-phasic foot pad secretion prevents slipping. J R Soc Interface 7:587–593
Dixon AFG, Croghan PC, Gowing RP (1990) The mechanism by which aphids adhere to smooth surfaces. J Exp Biol 152:243–253
Drechsler P, Federle W (2006) Biomechanics of smooth adhesive pads in insects: influence of tarsal secretion on attachment performance. J Comp Physiol A 192(11):1213–1222
Edwards JS, Tarkanian M (1970) The adhesive pads of Heteroptera: a re-examination. Proc R Ent Soc Lond A 45:1–5
Federle W (2006) Why are so many adhesive pads hairy? J Exp Biol 209(14):2611–2621
Gorb S (2001) Attachment devices of insect cuticle. Springer Science & Business Media, Dordrecht/Boston/London, 305pp
Gorb S, Gorb E, Kastner V (2001) Scale effects on the attachment pads and friction forces in syrphid flies (Diptera, Syrphidae). J Exp Biol 204(8):1421–1431
Gorb S, Varenberg M, Peressadko A, Tuma J (2007) Biomimetic mushroom-shaped fibrillar adhesive microstructure. J R Soc Interface 4(13):271–275
Gorb E, Hosoda N, Miksch C, Gorb S (2010) Slippery pores: anti-adhesive effect of nanoporous substrates on the beetle attachment system. J R Soc Interface 7(52):1571–1579
Homann H (1957) Haften Spinnen an einer Wasserhaut? Naturwissenschaften 44(11):318–319
Hsu PY, Ge L, Li X, Stark AY, Wesdemiotis C, Niewiarowski PH, Dhinojwala A (2011) Direct evidence of phospholipids in gecko footprints and spatula–substrate contact interface detected using surface-sensitive spectroscopy. J R Soc Interface 9(69):657–664
Huber G, Mantz H, Spolenak R, Mecke K, Jacobs K, Gorb SN, Arzt E (2005) Evidence for capillarity contributions to gecko adhesion from single spatula nanomechanical measurements. Proc Natl Acad Sci U S A 102(45):16293–16296
Israelachvili JN (1991) Intermolecular and surface forces, 2nd edn. Academic, London/San Diego, 450 pp
Jang J, Schatz GC, Ratner MA (2004) Capillary force in atomic force microscopy. J Chem Phys 120(3):1157–1160
Jiao Y, Gorb SN, Scherge M (2000) Adhesion measured on the attachment pads of *Tettigonia viridissima* (Orthoptera, Insecta). J Exp Biol 203:1887–1895
Kesel A, Martin A, Seidl T (2003) Adhesion measurements on the attachment devices of the jumping spider *Evarcha arcuata*. J Exp Biol 206(16):2733–2738
Kropf C, Bauer D, Schläppi T, Jacob A (2012) An organic coating keeps orb-weaving spiders (Araneae, Araneoidea, Araneidae) from sticking to their own capture threads. J Zool Syst Evol Res 50(1):14–18

Labonte D, Federle W (2015a) Rate-dependence of 'wet' biological adhesives and the function of the pad secretion in insects. Soft Matter 11(44):8661–8673
Labonte D, Federle W (2015b) Scaling and biomechanics of surface attachment in climbing animals. Phil Trans R Soc B 370(1661):20140027
Labonte D, Clemente CJ, Dittrich A, Kuo C-Y, Crosby AJ, Irschick DJ, Federle W (2016) Extreme positive allometry of animal adhesive pads and the size limits of adhesion-based climbing. Proc Natl Acad Sci U S A 113(5):1297–1302
Mizutani K, Egashira K, Toukai T, Ogushi J (2006) Adhesive force of a spider mite, *Tetranychus urticae*, to a flat smooth surface. JSME Int J Ser C 49(2):539–544
Niederegger S, Gorb SN (2006) Friction and adhesion in the tarsal and metatarsal scopulae of spiders. J Comp Physiol A 192(11):1223–1232
Niewiarowski PH, Lopez S, Ge L, Hagan E, Dhinojwala A (2008) Sticky gecko feet: the role of temperature and humidity. Plos One 3(5):e2192
Peattie A, Full R (2007) Phylogenetic analysis of the scaling of wet and dry biological fibrillar adhesives. Proc Natl Acad Sci U S A 104(47):18595–18600
Peattie AM, Dirks J-H, Henriques S, Federle W (2011) Arachnids secrete a fluid over their adhesive pads. Plos One 6(5), e20485
Peisker H, Michels J, Gorb SN (2013) Evidence for a material gradient in the adhesive tarsal setae of the ladybird beetle *Coccinella septempunctata*. Nat Comm 4:1661
Peressadko A, Gorb SN (2004) When less is more: experimental evidence for tenacity enhancement by division of contact area. J Adh 80(4):247–261
Prowse MS, Wilkinson M, Puthoff JB, Mayer G, Autumn K (2011) Effects of humidity on the mechanical properties of gecko setae. Acta Biomater 7(2):733–738
Roscoe D, Walker G (1991) Adhesion of spiders to smooth surfaces. Bull Br Arachnol Soc 8:224–226
Scherge M, Gorb SN (2001) Biological micro- and nano-tribology. Springer, Berlin
Stork NE (1980) Experimental analysis of adhesion of *Chrysolina polita* (Chrysomelidae, Coleoptera) on a variety of surfaces. J Exp Biol 88:91–107
Varenberg M, Pugno NM, Gorb SN (2010) Spatulate structures in biological fibrillar adhesion. Soft Matter 6:3269–3272
Varenberg M, Murarash B, Kligerman Y, Gorb SN (2011) Geometry-controlled adhesion: revisiting the contact splitting hypothesis. Appl Phys A 103:933–938
Wohlfart E (2005) Kraftmessungen am Haftsystem der Spinnenbeine von *Cupiennius salei* Keyserling (Arachnida, Ctenidae). Diploma thesis. University of Stuttgart, Stuttgart
Wolff JO, Gorb SN (2012) The influence of humidity on the attachment ability of the spider *Philodromus dispar* (Araneae, Philodromidae). Proc R Soc B 279(1726):139–143
Wolff JO, Gorb SN (2013) Radial arrangement of Janus-like setae permits friction control in spiders. Sci Rep 3:1101
Wolff JO, Seiter M, Gorb SN (2015) Functional anatomy of the pretarsus in whip spiders (Arachnida, Amblypygi). Arthropod Struct Dev 44(6):524–540

Chapter 11
Biomimetics: What Can We Learn From Arachnids?

Abstract In this chapter we discuss features rendering arachnids interesting models for the development of biomimetic adhesive materials and attachment systems. Geckoes have been established as a model for the development of dry adhesive materials. Spiders and trombidiform mites exhibit very similar adhesive structures to those of geckoes, and can be much easier maintained in the lab and manipulated in experiments. Complex hierarchical tenent setae are still not technically reproducible in all essential details, and existing 'gecko-inspired' adhesives are still very limited in their performance in the real world applications. The alternative structure of the spatulate adhesive surfaces of whip-spider feet may represent a new source of inspiration. We further present some unique principles recently explored in arachnids that may inspire the next generation of biomimetic adhesive materials, gripping devices, and glue applications. These are (1) the smart glue application in spider silk anchors, (2) the adhesive nano-fibre-mats of cribellar capture threads, (3) the prehensile feet of harvestmen, (4) the transverse stripe-shaped contact of some arolia, (5) the pressure sensitive adhesives of spider capture threads and harvestmen, (6) the droplet retaining hairs of harvestmen, and (7) the miniature suction cups of mites.

11.1 Why Geckoes Are Not Unique

Studying the functional morphology of attachment devices helps to understand the biology of organisms and their interactions with their environment, but is also of direct practical relevance. For instance, a basic understanding of the function of attachment devices in parasites and pests, represented by many ticks and mites, is crucial for the development of efficient counter-measures, such as tick-safe hiking clothing. Additionally, we can also utilize the results arising from such basic biological research, to gain new technical innovations by using various creative approaches.

In recent times, biodiversity has been recognized as an important source of inspiration for materials, surfaces, drugs, processing technologies, and other innovations. Most of the structures and phenomena, we observe in nature, are products of long-term evolution and represent efficient solutions for various problems. Since such

J.O. Wolff, S.N. Gorb, *Attachment Structures and Adhesive Secretions in Arachnids*, Biologically-Inspired Systems 7,
DOI 10.1007/978-3-319-45713-0_11

structures have evolved under the limitation of resources, they also often represent highly economic and sustainable solutions. There are two major approaches to transfer biological, biochemical or biophysical findings into technological applications. First, by studying the relationship between structure and function, we can get inspired for the development of novel high-end materials and devices. This approach is called bioinspiration or biomimetics. Intriguing examples of biological models which were used in such approaches are the burrs of burdock (Velcro fasteners), the wing tips of birds (winglets of modern aircrafts), the self-cleaning surfaces of plants (Lotus paint), and the drag-reducing skin of sharks (riblet foils) (Nachtigall 1998; Hesselberg 2007; Solga et al. 2007; Bhushan 2009; Dean and Bhushan 2010), to name a few. The second type of approach is the identification and isolation of the functional chemical components of biological materials and their up-scaled production by means of biotechnological techniques (Hennebert et al. 2015). The silk fibres of spiders (Heim et al. 2009; Porter and Vollrath 2009) and the adhesive proteins of mussels and barnacles (Deming 1999; Wiegemann 2005; Kamino 2008) are in the focus of this kind of research.

One industrial area that may highly benefit from such approaches is adhesives technology (Flammang and Santos 2015). Glues and adhesives have advanced to one of the most important bonding technologies in recent times, thanks to their versatility in application and the development of tailored high-end glues achieving high bonding strength and durability (Skeist 2012; Kinloch 2012). Yet there are many drawbacks of modern adhesives, such as the lack of reusability, susceptibility to humidity and UV radiation, limited toughness and inner strength, the long duration and/or complicated procedures to stabilize the bonding, dependence on hazardous and non-sustainable substances, high costs, or, in the case of medical applications, a lack of biocompatibility. However, many biological adhesives exhibit a set of desirable properties, presenting solutions for some of these problems.

One phenomenon that has been fascinated people over hundreds of years is the ability of some animals to scale smooth vertical or even inverted surfaces. This is possible due to a combination of friction and adhesion with rapid attachment-detachment cycles, which cannot be performed by using any conventional artificial adhesives. Since the discovery of the fact that many locomotory attachment pads are based on a specific micro-structure of their surfaces, there were numerous attempts to mimic this principle by the fabrication of micro-structured polymer tapes (Boesel et al. 2010). The toe pads of geckoes are in the focus of this research, mainly for one reason: the Tokey gecko is the largest animal with structural dry adhesive pads, hence it is closest to larger scaled technical applications. Furthermore, vertebrates are more appealing objects to most people, and research on them usually gains higher visibility and attention from the publicity than related research on arthropods. However, the adhesive setae of geckoes have a very delicate and complex structure, which cannot be reasonably reproduced by any fabrication method to date (Niewiarowski et al. 2016). Over years, the development of gecko-inspired adhesives with comparable properties as in its natural model was stuck by sticking to the gecko model. Few research groups used an explorative biodiversity approach, studying different organisms with similar adhesive abilities, to extract universal structural principles and to find novel unconventional structures with a similar function, which could be more easily reproduced (Gorb 2001; Arzt et al. 2003; Federle 2006; Gorb 2008). Eventually, a tape based on mushroom-like microstructures, inspired by the setae of male leaf beetles that use them to attach themselves to the smooth elytra of the mating partner, made the run by being the first commercially available, re-usable and versatile dry adhesive tape (Gorb et al. 2007). However, this tape is not switchable between high and low adhesion states and is susceptible to fine roughness of the substrate (Kasem and Varenberg 2013). As discussed in Chap. 5, mushroom-like microstructures may ideally serve longer term attachment, in contrast to the spatula like structures which are most efficient during dynamic attachment (Gorb and Varenberg 2007). This means that when we consequently follow the bio-inspiration approach, one has to bear in mind the requirements of the aimed application, e.g. whether durable and strong or highly dynamic adhesion is aimed. Accordingly, the natural history of the animals and the biological function of their structures cannot be ignored. Since adhesives are used for various applications with very different demands, it is worthwhile keeping an eye on nature to find novel solutions for the development of specific adhesives with tailored properties.

Arachnids can be an enormous source of inspiration for future innovations in the field of adhesives and bonding technologies. The perhaps most acknowledged example of spider-inspired technology is the fabrication of spider silk based superfibres (Heidebrecht et al. 2015). Additionally, there is recently increasing awareness that arachnids produce high-performance adhesives with a high potential for technical innovations, leading to biomimetic studies on the adhesive pads of jumping spiders (Seidl and Vidoni 2013), the glue-coated capture threads of orb web spiders (Sahni et al. 2011b) and spider silk thread anchors (Jain et al. 2014).

Arachnids as models for studies of bioadhesion are almost ideal objects. They can easily be maintained in the lab in great numbers and manipulated in experiments. There are no ethical concerns and restrictions as it is the case for vertebrates in many countries. In the case of gecko-like dry adhesives, there are very similar structures in hunting spiders and predatory mites, as depicted in Chap. 3. On the other hand, there are many unique structures and organs, which were not found elsewhere. This book may be a source of inspiration fostering future research by showing a broad range of adhesives and attachment devices in arachnids. Below some examples that we think are of high potential for biomimetics.

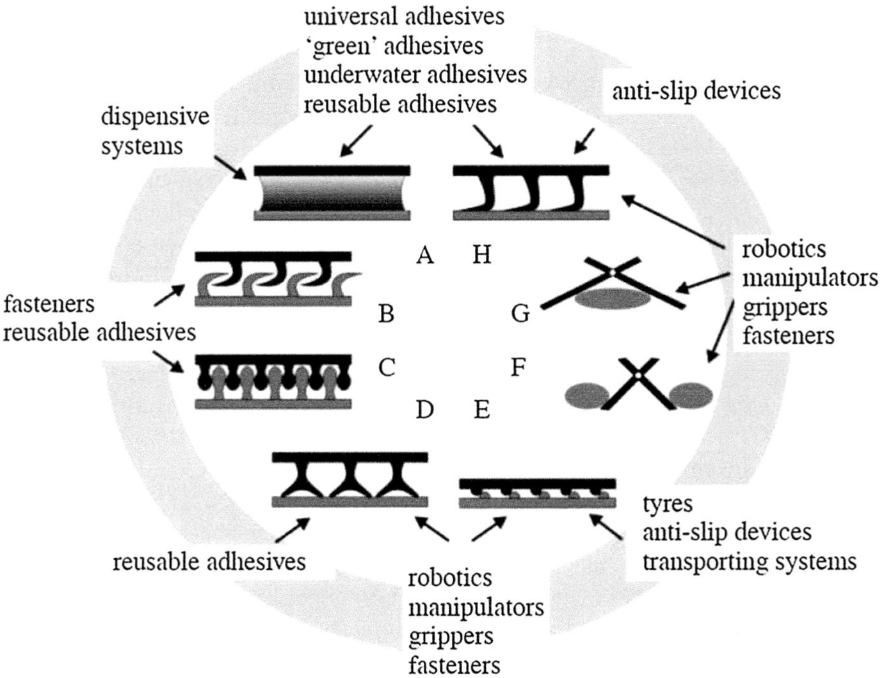

Fig. 11.1 Areas of existing and potential biomimetic applications based on studies of biological attachment devices. *A*, wet adhesion; *B*, hooks; *C*, lock; *D*, suction cups; *E*, friction; *F*, spacer; *G*, clamp; *H*, dry adhesion (From Gorb 2008)

11.2 Noteworthy Structures and Their Potential for Biomimetics and Bioinspiration

11.2.1 Micro-Patterned Surface of the Whip-Spider Arolium

The development of micro patterned dry adhesive tapes is the primary aim of biomimetic studies on the hairy adhesive pads of geckoes, insects, and spiders. The main problems that limit the direct transfer of found structural principles into technical solutions are (1) different application demands, (2) the upscaling to the relevant sizes in macroscopic applications, and (3) the restrictions in fabrication processes. The most successful innovation is an adhesive tape based on mushroom-shaped microstructures inspired by the modified tenent setae of male leaf beetles (Daltorio et al. 2005; Kim and Sitti 2006; Gorb et al. 2007). This material can produce high adhesive forces, when the force acts perpendicular, but is sensitive to the substrate roughness (Kasem and Varenberg 2013), and pull-off angles below 45° (Heepe et al. 2014). Because the microstructures bend at the surface during sliding, it is not useful for generation of strong friction (Varenberg and Gorb 2007). Spatulate structures have been shown to be more effective at low pull-off angles (Autumn et al. 2000) and can generate anisotropic high friction forces when sheared over smooth surfaces (Autumn et al. 2006; Niederegger and Gorb 2006; Wolff and Gorb 2013). Further, they can better adapt to rough surfaces. The main challenge of the implementation of this principle in biomimetic adhesive surfaces is the fabrication of flexible pillar structures with spatulate tips, which is hard to achieve at microscale.

The spatulate hexagonally-arranged structures found in amblypygids are unique among attachment devices in animals studied so far. We found, that they can produce high adhesive forces and may also have a drainage effect, which enhances their efficacy on wet surfaces. Amblypygid-inspired microstructures may be easier to produce using polymers than gecko-like ones. For instance, such structures could be reproduced by molding replica from the natural model, a technique that has previously been applied on super-hydrophobic surfaces of plants and arthropods (Koch and Barthlott 2009; Hensel et al. 2013). They could be useful in applications that demand high friction generation on smooth and wet surfaces.

11.2.2 Adhesive Nano-fibre Bundles of Cribellar Spider Silk

The cribellar capture threads of spiders are structural adhesives totally different from the adhesive structures of locomotory attachment pads, which are the primary source of inspiration for the development of fibrillar structural adhesives. Instead of complex microstructures, the cribellar thread is simply based on the bunch of loosely-packed nano-fibres. Nowadays fibres of comparable submicron diameters can be easily produced by electrospinning (Grafe and Graham 2003; Brown et al. 2016), allowing the fabrication of another kind of bioinspired fibrillar adhesives. The diameter, structure and arrangement of the nano-fibres in the cribellar thread play an important role and should be carefully considered prior to fabrication.

Also the method of spinning of silk fibres with diameters below 100 nm, as it is observed in the cribellar silk and in the silk of spider mites might be inspiring for textile research, because the conventional way of the production of synthetic fibres by extrusion through spinnerets does not allow formation of fibres with submicron diameter due to rheological constraints (Grafe and Graham 2003; Brown et al. 2016). The recently established way to produce nano-fibres by electrostatic pulling forces (electrospinning) has some practical limitations, such as the necessity of the use of both high voltage and a conductive collector, and difficulties to control the diameter and application pattern of fibres. The understanding of how spiders and mites manage to produce such delicate fibres by spinning under ambient conditions might help to improve artificial spinning techniques.

11.2.3 Transverse Lip Arolium

A transverse strip-like contact area of the arolium seems to be a special feature of arachnids. Such geometry of smooth adhesive pads is not common in insects. In arachnids, it has been independently evolved three times: in whip spiders, pseudoscorpions, and anystid mites. This shape may permit a high anisotropy of friction and adhesion and enable a rapid switch between stress distribution, when forces act towards the body, and stress concentration during lateral peeling. This principle could be used for the design of adhesive grippers, e.g. for microfabrication techniques, or in robotics. In order to transport large sheets or panes, suction cups are usually used, which, however, may deform the material due to the generated local pressure difference. Soft adhesive pads may produce much lower stress in the contact zone with the surface of transported material. Strip-like pads could carry components with smooth surfaces by generating high friction in the direction of the long side, and release them by a transverse peeling mechanism along the short side.

11.2.4 Prehensile Feet of Harvestmen

The tarsal flexor system of the long-legged harvestmen already fascinated scientists more than 80 years ago, because its control mechanism is simple, but the efficacy is high (Kästner 1931). In insects, solifuges and scorpions, the tarsi are also subsegmented, but never in such an excessive extend that they can be used as prehensile organs. Furthermore, in contrast to harvestmen, the insect tarsi bend upwards. Hence, the prehensile tarsi of harvestmen represent a unique system, which can effectively attach to various substrates without having elaborate adhesive organs. Instead, friction is generated by an array of simple pointed setae with the tips at nanoscale. This principle might be interesting for the design of grippers for handling long objects, like cables, poles or pillars. It might also be used, for example, for tree climbing equipment.

11.2.5 Spider SilkAnchors

Spiders use glue-coated fibres in a remarkable and unique way to create strong thread attachments (Sahni et al. 2012; Pugno et al. 2013; Wolff et al. 2015). In artificial materials, the insertion of flexible fibres is known as an effective means to increase the toughness of brittle materials (Li 2003). Artificial adhesives are still limited in their toughness, and bondings often fail due to internal fracture (cohesion failure). Embedded micro-fibres, like in the silk anchor glue, may enhance the extensibility of adhesives and serve as effective crack stoppers.

The hierarchical structure of attachment discs may further teach us how to apply a limited amount of glue in an optimal manner, to achieve the highest bonding strength, even on hardly wettable surfaces. Since many tailored high-end glues are very expensive, this may be an important means of enhancing cost-efficiency and reducing resource consumption. First steps in this direction have been made by using an electrospinning technique (Jain et al. 2014).

Future research on the exact chemical composition of the piriform silk glue and its material alteration during the quick drying process may lead to novel sustainable and strong bio-adhesives. For this purpose a chemical and further structural analysis are of high importance. Biological glues are especially of interest in medicine for the permanent or temporary gluing of bone fractures, tissues, and implants.

Further, the investigation of spider spigots could also lead to innovations that enhance the efficacy of glue nozzles. Both the piriform and the aggregate gland spigots are usually remarkably clean, in spite of adhesive fluids, coming out of them. Adhering remains of dried secretion, which could plug the gland opening, are very rare. Therefore, further studies of structural, chemical and physical principles, contributing to the prevention of spigot contamination by the glue, are very promising.

11.2.6 Pressure Sensitive Adhesives in Capture Threads of Spiders and Harvestmen

The viscid glue on the capture threads of spiders is a re-usable adhesive with durable, strong stickiness that, in contrast to synthetic adhesives, does not show reduced, but enhanced stickiness under humid conditions. It is regarded as an important source of inspiration to improve artificial pressure sensitive adhesives, especially for their use in humid environments (Sahni et al. 2010, 2011a, b; Amarpuri et al. 2015).

Similarly, the fast spreading and good adhesion of nemastomatid glue even to the Lotus-leaf-like surfaces of springtails is remarkable. The fluid switches between low and high viscosity extremely quickly. Studying the molecular composition of the fluid may provide insights into mechanisms that can improve current pressure sensitive adhesives, especially in a dynamic kind of applications, and enhance wettability of glues or paints on micro-structured and fluid-repellent surfaces.

11.2.7 Droplet Bearing Setae of Harvestmen

The anchorage of a fluid glue droplet to secretory setae covered by tiny microtrichia independently evolved in Palpatores harvestmen and mesostigmatid mites. This principle could be interesting for droplet catching devices, e.g. for gluing with minute amounts of the adhesive in microfabrication, or for micro-/nanoprinting technologies.

11.2.8 Minute Suction Cups

Among reversible adhesive structures, suckers can generate much higher attachment forces than smooth and hairy pads, but this is only true at the macro scale (Spolenak et al. 2005). Previous theoretical models indicated that the function and thus evolution of suction cups with a diameter smaller than 10 μm is unlikely (Spolenak et al. 2005). Keeping this in mind, the occurrence of minute suction cups in terrestrial acaridid mites is quite astonishing. Studies on the material properties and biomechanics of these minute suction cups could bring new insights in potential means of efficacy enhancement of suction cups at a small scale, which might be of high interest for micro-fabrication and micro-robotics. The great advantage of such micro-suckers in contrast to stickers is the switchability of adhesion by the elevation or depression of the sucker plate.

References

Amarpuri G, Chaurasia V, Jain D, Blackledge TA, Dhinojwala A (2015) Ubiquitous distribution of salts and proteins in spider glue enhances spider silk adhesion. Sci Rep 5:9030

Arzt E, Gorb S, Spolenak R (2003) From micro to nano contacts in biological attachment devices. Proc Natl Acad Sci U S A 100(19):10603–10606

Autumn K, Liang YA, Hsieh ST, Zesch W, Chan WP, Kenny TW, Fearing R, Full RJ (2000) Adhesive force of a single gecko foot-hair. Nature 405:681–685

Autumn K, Dittmore A, Santos D, Spenko M, Cutkosky M (2006) Frictional adhesion: a new angle on gecko attachment. J Exp Biol 209:3569–3579

Bhushan B (2009) Biomimetics: lessons from nature – an overview. Phil Trans R Soc A 367(1893):1445–1486

Boesel LF, Greiner C, Arzt E, del Campo A (2010) Gecko-inspired surfaces: a path to strong and reversible dry adhesives. Adv Mater 22(19):2125–2137

Brown TD, Dalton PD, Hutmacher DW (2016) Melt electrospinning today: an opportune time for an emerging polymer process. Prog Polym Sci 56:116–166

Daltorio KA, Horchler AD, Gorb SN, Ritzmann RE, Quinn RD (2005) A small wall-walking robot with compliant, adhesive feet. In: Proceedings of IEEE international conference on intelligent robots and systems. Edmonton, Canada, pp 4018–4023

Dean B, Bhushan B (2010) Shark-skin surfaces for fluid-drag reduction in turbulent flow: a review. Phil Trans R Soc Lond 368(1929):4775–4806

Deming TJ (1999) Mussel byssus and biomolecular materials. Curr Opin Chem Biol 3(1):100–105

Federle W (2006) Why are so many adhesive pads hairy? J Exp Biol 209(14):2611–2621

Flammang P, Santos R (2015) Biological adhesives: from biology to biomimetics. Interface Focus 5:20140086

Gorb S (2001) Attachment devices of insect cuticle. Springer Science & Business Media, Dordrecht/Boston/London, 305 pp

Gorb SN (2008) Biological attachment devices: exploring nature's diversity for biomimetics. Phil Trans R Soc 366(1870):1557–1574

Gorb S, Varenberg M (2007) Mushroom-shaped geometry of contact elements in biological adhesive systems. J Adhes Sci Technol 21:1175–1183

Gorb S, Varenberg M, Peressadko A, Tuma J (2007) Biomimetic mushroom-shaped fibrillar adhesive microstructure. J R Soc Interface 4(13):271–275

Grafe T, Graham K (2003) Polymeric nanofibers and nanofiber webs: a new class of nonwovens. Nonwoven Technol Rev 12:51–55

Heepe L, Carbone G, Pierro E, Kovalev AE, Gorb SN (2014) Adhesion tilt-tolerance in bio-inspired mushroom-shaped adhesive microstructure. Appl Phys Lett 104(1):011906

Heidebrecht A, Eisoldt L, Diehl J, Schmidt A, Geffers M, Lang G, Scheibel T (2015) Biomimetic fibers made of recombinant spidroins with the same toughness as natural spider silk. Adv Mater 27(13):2189–2194

Heim M, Keerl D, Scheibel T (2009) Spider silk: from soluble protein to extraordinary fiber. Angew Chem Int Ed 48(20):3584–3596

Hennebert E, Maldonado B, Ladurner P, Flammang P, Santos R (2015) Experimental strategies for the identification and characterization of adhesive proteins in animals: a review. Interface Focus 5(1):20140064

Hensel R, Helbig R, Aland S, Voigt A, Neinhuis C, Werner C (2013) Tunable nano-replication to explore the omniphobic characteristics of springtail skin. NPG Asia Mater 5(2), e37

Hesselberg T (2007) Biomimetics and the case of the remarkable ragworms. Naturwissenschaften 94(8):613–621

Jain D, Sahni V, Dhinojwala A (2014) Synthetic adhesive attachment discs inspired by spider's pyriform silk architecture. J Polym Sci B Polym Phys 52(8):553–560

Kamino K (2008) Underwater adhesive of marine organisms as the vital link between biological science and material science. Mar Biotechnol 10(2):111–121

Kasem H, Varenberg M (2013) Effect of counterface roughness on adhesion of mushroom-shaped microstructure. J R Soc Interface 10(87):20130620

Kästner A (1931) Biologische Beobachtungen an Phalangiiden. Zool Anz 95:293–302

Kim S, Sitti M (2006) Biologically inspired polymer microfibers with spatulate tips as repeatable fibrillar adhesives. Appl Phys Lett 89:261911

Kinloch A (2012) Adhesion and adhesives: science and technology. Springer Science & Business Media, Dordrecht

Koch K, Barthlott W (2009) Superhydrophobic and superhydrophilic plant surfaces: an inspiration for biomimetic materials. Phil Trans R Soc A 367:1487–1509

Li VC (2003) On engineered cementitious composites (ECC): a review of the material and its applications. J Adv Concr Technol 1(3):215–230

Nachtigall W (1998) Bionik—Was ist das? Springer Science & Business Media, Berlin, Heidelberg

Niederegger S, Gorb SN (2006) Friction and adhesion in the tarsal and metatarsal scopulae of spiders. J Comp Physiol A 192:1223–1232

Niewiarowski P, Stark AY, Dhinojwala A (2016) Sticking to the story: outstanding challenges in gecko-inspired adhesives. J Exp Biol 219(7):912–919

Porter D, Vollrath F (2009) Silk as a biomimetic ideal for structural polymers. Adv Mater 21(4):487–492

Pugno NM, Cranford SW, Buehler MJ (2013) Synergetic material and structure optimization yields robust spider web anchorages. Small 9(16):2747–2756

Sahni V, Blackledge TA, Dhinojwala A (2010) Viscoelastic solids explain spider web stickiness. Nat Commun 1:19

Sahni V, Blackledge TA, Dhinojwala A (2011a) A review on spider silk adhesion. J Adhes 87(6):595–614

Sahni V, Labhasetwar DV, Dhinojwala A (2011b) Spider silk inspired functional microthreads. Langmuir 28(4):2206–2210

Sahni V, Harris J, Blackledge TA, Dhinojwala A (2012) Cobweb-weaving spiders produce different attachment discs for locomotion and prey capture. Nat Comm 3:1106

Seidl T, Vidoni R (2013) Adhesion to flat surfaces: from spiders to stickers. In: Nentwig W (ed) Spider ecophysiology. Springer, Berlin/New York, pp 463–473

Skeist I (2012) Handbook of adhesives. Springer Science & Business Media, New York

Solga A, Cerman Z, Striffler BF, Spaeth M, Barthlott W (2007) The dream of staying clean: lotus and biomimetic surfaces. Bioinspir Biomim 2(4):S126

Spolenak R, Gorb S, Gao H, Arzt E (2005) Effects of contact shape on the scaling of biological attachments. Proc R Soc A 2054:305–319

Varenberg M, Gorb S (2007) Shearing of fibrillar adhesive microstructure: friction and shear-related changes in pull-off force. J R Soc Interface 4(15):721–725

Wiegemann M (2005) Adhesion in blue mussels (*Mytilus edulis*) and barnacles (genus *Balanus*): mechanisms and technical applications. Aquat Sci 67(2):166–176

Wolff JO, Gorb SN (2013) Radial arrangement of Janus-like setae permits friction control in spiders. Sci Rep 3:1101

Wolff JO, Grawe I, Wirth M, Karstedt A, Gorb SN (2015) Spider's super-glue: thread anchors are composite adhesives with synergistic hierarchical organization. Soft Matter 11(12):2394–2403

Appendix

A1. List of Supplementary Videos (Online)

Chapter 2

Video 2.1 Movement of pretarsus in the basal spider *Vinathela tonkinensis* (Liphistiidae, juvenile) walking on a sheet of paper. Recorded with 1000 fps and playback with 30 fps.

Video 2.2 Movement of the pretarsus of a hind leg in the haplogyne spider *Harpactea* sp. (Dysderidae) walking on a sheet of paper. Note that the claws are elevated prior the leg movement, enabling a proper detachment. Recorded with 1000 fps and playback with 30 fps.

Video 2.3 Movement of the pretarsus in a hind leg of the apulvillate whip-spider *Phrynus longipes* (Phrynidae) walking on a sheet of paper. Recorded with 500 fps and playback with 15 fps.

Video 2.4 Attachment, detachment and leg propulsion in the whip-scorpion *Typopeltis crucifer* (Thelyphonidae) walking on a wooden bar. Recorded with 500 fps and playback with 15 fps.

Video 2.5 Sac spider *Clubiona* sp. (Clubionidae) attacking a fruit fly. Note the elevation of leg spines during the attack, representing a switchable barrier suppressing prey escape. Also note the eventual grasping with the chelicerae, which work like clamps. Recorded with 2000 fps and playback with 30 fps. Video by Neele Ott, with kind permission.

Video 2.6 Lynx spider (Oxyopidae) capturing a fruit fly, using a full leg basket. In species of this family the long spines are durably erected. Recorded with 2000 fps and playback with 30 fps. Video by Neele Ott, with kind permission.

Video 2.7 Crab spider *Xysticus* sp. (Thomisidae) capturing a fly with its front leg. Recorded with 500 fps and playback with 20 fps. Video by Neele Ott, with kind permission.

J.O. Wolff, S.N. Gorb, *Attachment Structures and Adhesive Secretions in Arachnids*, Biologically-Inspired Systems 7,
DOI 10.1007/978-3-319-45713-0

Video 2.8 *Palpimanus gibbulus* (Palpimanidae) attacking a ghost spider (*Anyphaena accentuata*). These specialized spider-hunters have strong clamp-like front legs equipped with a dense scopula. Recorded with 2000 fps and playback with 30 fps. Video by Neele Ott, with kind permission.

Chapter 3

Video 3.1 Transmission light HSV of the foot of a *Zoropsis spinimana* (Zoropsidae), showing the spreading of the claw tuft. Recorded with 500 fps and playback with 25 fps.

Video 3.2 RICM-HSV of the claw tuft of a *Zoropsis spinimana* (Zoropsidae), showing the detachment of tenent microtrichia. In this species the claw tuft setae are not broadened. The setae in the right that stay in contact after claw tuft detachment are the frictional setae of the distal tarsus. Note minute traces left behind by the detaching tenent setae, which might be remains of a fluid secretion present on the spatulae. Recorded with 500 fps and playback with 15 fps.

Video 3.3 RICM-HSV of claw tuft setae of the ghost spider *Anyphaena accentuata* (Anyphaenidae), showing attachment and detachment of tenent microtrichia. This species exhibits very broad claw tuft setae. Recorded with 500 fps and playback with 15 fps.

Video 3.4 RICM-HSV of a claw tuft seta of the ghost spider *Anyphaena accentuata* (Anyphaenidae), showing detachment of tenent microtrichia. Recorded with 500 fps and playback with 15 fps.

Video 3.5 RICM-HSV of the claw tuft of an unidentified Palpimanidae from South-Africa, showing dry mode of adhesion (no traces of tenent setae left behind). Recorded with 500 fps and playback with 15 fps.

Video 3.6 RICM-HSV of the claw tuft of an unidentified Palpimanidae from South-Africa, showing wet mode of adhesion (fluid coating of claw tuft setae leaving droplets behind after detachment). Recorded with 500 fps and playback with 15 fps.

Video 3.7 RICM-HSV of the claw tuft of an unidentified Palpimanidae from South-Africa, showing semi-wet mode of adhesion (left claw tuft setae dry, and right claw tuft setae wet). Recorded with 500 fps and playback with 15 fps.

Video 3.8 RICM-HSV of claw tuft setae of an ablated leg of a *Cupiennius salei* spider (Ctenidae). In this experiment the contacting glass slide is sheared against the setae (proximal direction), causing detachment of spatulae.

Video 3.9 RICM-HSV of a foot of the snout mite *Cyta* sp. (Bdellidae), showing attachment and detachment of the tenent empodium. Recorded with 1000 fps and playback with 15 fps.

Chapter 6

Video 6.1 RICM-HSV of the adanal sucker plate of a phoretic deutonymph (hypopus) of the mite *Histiostoma* sp. (Histiostomatidae), showing movement of suckers, detachment and traces of viscous fluid left behind.

Video 6.2 RICM-HSV of the adanal sucker plate of a phoretic deutonymph (hypopus) of the mite *Histiostoma* sp. (Histiostomatidae), showing retraction and protraction of one of the main suckers.

Chapter 7

Video 7.1 Nymph of the scorpion *Liocheles australasiae* (Hemiscorpiidae) walking upside down on a glass coverslip. Playback with real speed.

Video 7.2 Detail of a leg tip (lateral view) of a *L. australasiae* nymph walking on a glass coverslip, showing the detachment of the arolium by invagination. Recorded with 50 fps and playback with 30 fps.

Video 7.3 Detail of a leg tip (lateral view) of a *L. australasiae* nymph walking on a glass coverslip, showing the detachment of the arolium by invagination. Recorded with 50 fps and playback with 15 fps.

Video 7.4 Detail of a leg tip (view from below) of a *L. australasiae* nymph walking on a glass coverslip, showing the detachment of the arolium by invagination. Recorded with 500 fps and playback with 30 fps.

Video 7.5 Detail of two leg tips (view from below) of a *L. australasiae* nymph walking on a glass coverslip, showing the detachment of the arolium by invagination (transmission illumination and DIC optic). Note internal features like the retractor tendon and hemolymph cells. Recorded with 500 fps and playback with 30 fps.

Video 7.6 RICM-HSV of the arolium of a *L. australasiae* nymph, showing dry mode of adhesion (no traces left behind after detachment). Recorded with 1000 fps and playback with 30 fps.

Video 7.7 RICM-HSV of the arolium of a *L. australasiae* nymph, showing wet mode of adhesion (droplets of a fluid secretion left behind after detachment). Note that the secretion keeps on shrinking to droplets on the hydrophilic glass, indicating their lipid-like nature. Also note the dirt particle, which does not stick to the arolium after detachment, but remains on the glass slide. Hence, the high amount of fluids observed here, may be part of a rinsing mechanism of the arolium. Recorded with 1000 fps and playback with 30 fps.

Video 7.8 Tick walking upside down on a glass slide. The tick was trapped between a common microscopy slide and the table to reduce its mobility. Nonetheless it was capable to push itself out of the glass slide, showing an enormous strength of the animal and high friction of its arolia. Note the inflation and folding of the arolia during attachment and detachment. Playback with real speed.

Video 7.9 Detail of a leg tip (view from above) of the whip-spider *Charon* cf. *grayi* (Charontidae) walking on a glass slide, showing the detachment of the arolium and movement of claws. Note the footprint left behind, indicating the deposition of fluid secretions. Recorded with 500 fps and playback with 30 fps.

Video 7.10 Detail of a leg tip (frontal view) of the whip-spider *Charon* cf. *grayi* (Charontidae) walking on a glass slide, showing the detachment of the arolium. Recorded with 3000 fps and playback with 30 fps.

Video 7.11 RICM-HSV of an arolium of the whip-spider *Charinuscubensis* (Charinidae), showing the contact behavior during detachment (dry mode of adhesion). Note that single spatulae jump independently out of contact, thus there is an uneven peel-off (stepwise crack propagation). Recorded with 6250 fps and playback with 15 fps.

Video 7.12 RICM-HSV of an arolium of the whip-spider *C.* cf. *grayi*, showing the contact behavior during detachment (wet mode of adhesion). Recorded with 3000 fps and playback with 15 fps.

Video 7.13 RICM-HSV of an arolium of the whip-spider *C.* cf. *grayi*, showing the contact behavior during detachment (partially wet mode of adhesion). Note that secretions occur only in single areas and do not spread over the pad. Recorded with 8000 fps and playback with 15 fps.

Video 7.14 RICM-HSV of an arolium of the whip-spider *Sarax brachydactylus* (Charinidae), showing the contact behavior during detachment (partially wet mode of adhesion). Note that a part of the secretions appears highly viscous, keeping the imprints of spatulae. Recorded with 8000 fps and playback with 25 fps.

Video 7.15 RICM-HSV of an arolium of the pseudoscorpion *Chthonius* sp. (Chthoniidae), showing detachment. Note the bi-lateral peel-off. Recorded with 5000 fps and playback with 30 fps.

Video 7.16 RICM-HSV of an arolium of the pseudoscorpion *Neobisium* sp. (Neobisidae), showing detachment. Note the bi-lateral peel-off. Recorded with 500 fps and playback with 30 fps.

Video 7.17 RICM-HSV of an arolium of the pseudoscorpion *Neobisium* sp., showing sliding of the arolium along the glass surface. Note that lipidoid fluids are deposited by the pointed tarsal setae. Recorded with 500 fps and playback with 30 fps.

Video 7.18 Tip of the walking leg of the predatory mite *Anystis* sp. (Anystidae), showing movements of the pretarsus. The foot exhibits both a pair of tenent setae and an arolium, and the high flexibility of the distal tarsus permits a stepwise contact of setae, arolium and claws. Recorded with 1000 fps and playback with 30 fps.

Video 7.19 RICM-HSV of the foot of *Anystis* sp. showing stepwise detachment of arolium and setae. Note the fluids left behind both by arolium and tenent setae. Recorded with 5400 fps and playback with 15 fps.

Video 7.20 RICM-HSV of the arolium of *Anystis* sp. showing its detachment. Note the bi-lateral peel-off and the lack of fluid remains. Recorded with 30000 fps and playback with 15 fps.
Video 7.21 RICM-HSV of the foot of *Anystis* sp. showing detachment and attachment. Note that in this case only the tenent setae are brought in contact. Recorded with 500 fps and playback with 30 fps.

Chapter 8

Video 8.1 The harvestman *Mitostoma chrysomelas* (Nemastomatidae) capturing a springtail with its sticky pedipalps. Real speed.
Video 8.2 *M. chrysomelas* capturing a springtail with its sticky pedipalps. Also note the action of the pincer-like chelicerae in the end. Recorded with 500 fps and playback with 15 fps.
Video 8.3 *M. chrysomelas* capturing a springtail with its sticky pedipalps. Recorded with 500 fps and playback with 15 fps.
Video 8.4 *M. chrysomelas* capturing a springtail by clamping its antenna between the tibia and the hyper-flexible tarsus of the sticky pedipalp. Recorded with 500 fps and playback with 15 fps.
Video 8.5 Demonstration of the stickiness and visco-elasticity of the glue on the pedipalp of *M. chrysomelas*. Real time.
Video 8.6 Adhesion test on a glue droplet of a pedipalpal glandular seta of *Nemastoma lugubre* (Nemastomatidae). The pedipalp is retracted from the glass beam by a motor at a rate of 1 mm/s. Recorded with 500 fps and playback with 30 fps.

A2. List of Studied Species

The following list compiles the species that were used in studies of which original data is presented throughout this book. Information from previous studies (i.e. Wolff et al. 2013 PLoS One) and meta-analysis is not included here.

Abbreviations:

Material used	liv, freshly collected or obtained from breeding stocks; pm, preserved material from museal or private collections
Parts studied	an, animal; se, secretion products only.
Experiments performed	fm, force measurements; his, histology; hsv, high speed-videography; lm, conventional light microscopy (macro

photography, reflection or transmission light microscopy); ricm, reflection-interference contrast microscopy (including ricm-hsv); sem, scanning electron microscopy.

Araneae

Mesothelae:

Liphistius birmanicus (Liphistiidae) liv; se; lm, sem
Liphistius ornatus (Liphistiidae) liv; se; lm, sem
Vinathela tonkinensis (Liphistiidae) liv; an; hsv, sem

Mygalomorphae:

Atypus piceus (Atypidae) liv; an, se; hsv, lm, sem
Cteniza sp. of Sardinia (Ctenizidae) liv; se; lm, sem
Idioctis sp. of the Philippines (Barychelidae) liv; se; lm, sem
Macrothele sp. from Turkey (Hexathelidae) liv; se; lm, sem
Aphonopelma seemanni (Theraphosidae) liv; an; sem

Araneomorphae:

Palpimanus gibbulus (Palpimanidae) liv; an; hsv, lm, sem
Undetermined sp. from South Africa (Palpimanidae) liv; an; lm, ricm, sem
Harpactea sp. (Dysderidae) liv; an; hsv, lm, sem
Pholcus phalangioides (Pholcidae) liv; an, se; hsv, lm, sem
Scytodes thoracica (Scytodidae) liv; an; hsv, sem
Loxosceles rufescens (Sicariidae) liv; se; lm
Cupiennius salei (Ctenidae) liv; an, se; fm, hsv, lm, ricm, sem
Phoneutria fera (Ctenidae) liv; an; lm, sem
Zoropsis spinimana (Zoropsidae) liv; an; fm, hsv, lm, ricm, sem
Xysticus sp. (Thomisidae) liv; an; hsv, sem
Anyphaena accentuata (Anyphaenidae) liv; an, sem
Clubiona reclusa (Clubionidae) liv; an, sem
Clubiona sp. (Clubionidae) liv; an; hsv, sem
Scotophaeus sp. from Northern Germany (Gnaphosidae) liv; an, se; lm, sem
Euophrys frontalis (Salticidae) liv; an; sem
Larinioides sclopetarius (Araneidae) liv; an, se; lm, sem
Nephila senegalensis (Araneidae) liv; an, se; lm, sem
Parasteatoda tepidariorum (Theridiidae) liv; se; lm

Amblypygi

Apulvillata:

Phrynus longipes (Phrynidae) liv; an, se; hsv, lm, sem

Pulvillata:

Charinus cubensis (Charinidae) liv; an; fm, his, hsv, lm, ricm, sem
Sarax brachydactylus (Charinidae) liv; an; fm, lm, ricm, sem
Charon cf. *grayi* (Charontidae) liv, pm; an, se; fm, hsv, lm, ricm, sem

Uropygi
Thelyphonida:
Typopeltis crucifer (Thelyphonidae) liv; an; his, hsv, lm, sem
Thelyphonus caudatus (Thelyphonidae) pm; an; lm

Schizomida:
Surazomus sturmi (Hubbardiidae) pm; an; lm
Undetermined sp. from Philippines liv; an; lm, sem

Scorpiones
Liocheles australasiae (Hemiscorpiidae) liv; an; fm, hsv, lm, ricm, sem
Euscorpius sp. (Euscorpiidae) liv; an; lm
Scorpiops sp. (Euscorpiidae) liv; an; lm, ricm

Solifugae
Undetermined sp. from South Africa (Solpugidae) liv; an; lm, sem

Pseudoscorpiones
Atemnus politus (Atemnidae) pm; an; lm, sem
Atemnus sp. (Atemnidae) liv; an; ricm, sem
Chthonius sp. from the Southern Alps (Chthoniidae) liv; an; ricm, sem
Neobisium sp. from Northern Germany (Neobisiidae) liv; an; fm, his, hsv, lm, ricm, sem
Neobisium spelaeum (Neobisiidae) pm; an; lm, sem
Roncus lubricus (Neobisiidae) pm; an; lm
Megachernes himalayensis (Chernetidae) pm; an; lm, sem
Pselaphochernes scorpioides (Chernetidae) pm; an; lm, sem
Lasiochernes pilosus (Chernetidae) pm; an; lm
Undetermined sp. from Philippines (Garypidae) liv; an; lm, sem
Geogarypus nigrimanus (Geogarypidae) pm; an; lm
Olpium pallipes (Olpiidae) pm; an; lm
Withius piger (Withiidae) pm; an; lm

Opiliones
Cyphophthalmi:
Meghalaya sp. (Stylocellidae) liv; an; lm, sem

Laniatores:
Paranonychus sp. (Triaenonychidae) pm; an; lm; sem
Holoscotolemon naturae (Cladonychiidae) liv; an; lm, ricm, sem
Holoscotolemon jaqueti (Cladonychiidae) liv; an; lm, sem
Holoscotolemon querilhaci (Cladonychiidae) pm; an; lm; sem
Gnomulus sp. from the Philippines (Sandokanidae) liv; an; lm, sem
Scotolemon doriae (Phalangodidae) pm; an; lm, sem
Scotolemon lucasi (Phalangodidae) pm; an; lm, sem
Sitalcina californica (Phalangodidae) pm; an; lm, sem
Dibunus similis (Epedanidae) pm; an; lm, sem
Cynortellana quadrimaculata (Cosmetidae) pm; an; lm, sem

Protimesius gracilis (Stygnidae) pm; an; lm
Metacrobunus sp. (Epedanidae) pm; an; lm, sem
Harpagonellus glaber (Epedanidae) pm; an; lm
Galibrotus cf. *riedeli* (Biantidae) pm; an; lm, sem
Hinzuanius flaviventris (Biantidae) pm; an; lm, sem
Biantes thakkali (Biantidae) pm; an; lm
Pellobunus insulcatus (Samoidae) pm; an; lm
Waigeucola palpalis (Samoidae) pm; an; lm
Undetermined sp. (Assamidae) pm; an; lm, sem

Eupnoi:
Dicranopalpus ramosus (Phalangiidae) liv; an; fm, lm, sem
Paroligolophus agrestis (Phalangiidae) liv; an; fm, lm, sem
Phalangium opilio (Phalangiidae) liv; an; lm, sem
Zacheus crista (Phalangiidae) pm; an; sem
Metagagrella cf. *minax* (Sclerosomatidae) pm; an; sem
Leiobunum blackwalli (Sclerosomatidae) liv; an; fm, hsv; lm; sem
Thrasychirus gulosus (Neopilionidae) pm; an; sem

Dyspnoi:
Sabacon simoni (Sabaconidae) liv; an; hsv, lm, sem
Trogulus martensi (Trogulidae) liv, pm; an; lm, sem
Dicranolasma sp. of the Southern Alps (Dicranolasmatidae) liv; an; lm, sem
Mitostoma chrysomelas (Nemastomatidae) liv; an; hsv, lm, sem

Acari
Opilioacariformes:
Opilioacarus texanus (Opilioacaridae) pm; an; sem

Ixodida:
Ixodes ricinus (Ixodidae) liv; an; his, hsv, lm, ricm, sem

Mesostigmata:
Amblyseius swirskii (Phytoseiidae) pm; an, sem
Uropoda orbicularis (Uropodidae) liv; se; sem
Undetermined sp. from Northern Germany liv; an; sem

Trombidiformes:
Caeculus sp. from the Southern Alps (Caeculidae) liv; an; sem
Anystis sp. from Northern Germany (Anystidae) liv; an; hsv, lm, ricm, sem
Abrolophus sp. from Northern Germany (Erythraeidae) liv; an; lm, ricm, sem
Cyta sp. from Northern Germany (Bdellidae) liv; an; lm, ricm, sem
Bryobia sp. from Northern Germany (Tetranychiidae) liv; an; sem
Linopodes sp. from Northern Germany (Cocceupodidae) liv; an; lm, ricm, sem

Sarcoptiformes:
Histiostoma sp. (Histiostomatidae) liv; an; ricm, sem
Undetermined sp. of Oribatidae from Northern Germany liv; an; sem

Ricinulei
Pseudocellus pachysoma (Ricinoididae) pm; an; lm; sem

Index

J.O. Wolff, S.N. Gorb, *Attachment Structures and Adhesive Secretions in Arachnids*, Biologically-Inspired Systems 7,
DOI 10.1007/978-3-319-45713-0

Printed by Printforce, the Netherlands